Sleep-Related Breathing Disorders

Advances in Oto-Rhino-Laryngology

Vol. 80

Series Editor

Patrick J. Bradley Nottingham

Sleep-Related Breathing Disorders

Volume Editor

Hsin-Ching Lin Kaohsiung

45 figures, 19 in color, and 14 tables, 2017

KARGER

Basel · Freiburg · Paris · London · New York · Chennai · New Delhi · Bangkok · Beijing · Shanghai · Tokyo · Kuala Lumpur · Singapore · Sydney

Hsin-Ching Lin
Department of Otolaryngology, Sleep Center
Kaohsiung Chang Gung Memorial Hospital
123, Ta-Pei Road, Niao-Sung District
Kaohsiung City 833, Taiwan (ROC)

Library of Congress Cataloging-in-Publication Data

Names: Lin, Hsin-Ching, editor.
Title: Sleep-related breathing disorders / volume editor, Hsin-Ching Lin.
Other titles: Sleep-related breathing disorders (Lin) | Advances in oto-rhino-laryngology ; v. 80. 0254-5020
Description: Basel : New York : Karger, 2017. | Series: Advances in oto-rhino-laryngology, ISSN 0065-3071; vol. 80 | Includes bibliographical references and indexes.
Identifiers: LCCN 2017026084| ISBN 9783318060645 (hard cover : alk. paper) | ISBN 9783318060652 (electronic version)
Subjects: | MESH: Sleep Apnea Syndromes--surgery
Classification: LCC RC737.5 | NLM WF 143 | DDC 616.2/09--dc23 LC record available at https://lccn.loc.gov/2017026084

Bibliographic Indices. This publication is listed in bibliographic services, including Current Contents® and Index Medicus.

www.karger.com
Printed on acid-free and non-aging paper (ISO 9706)
ISSN 0065–3071
e-ISSN 1662–2847
ISBN 978–3–318–06064–5
e-ISBN 978–3–318–06065–2

Contents

Preface

Sleep medicine is a relatively young medical specialty in the history of medicine. However, sleep has such a huge impact on our lives, both while sleeping and awake. Sleep-related breathing disorders, especially obstructive sleep apnea, obviously originate from upper airway abnormalities. The diagnosis, treatment, and research of these disorders includes multidisciplinary specialties. Significant advances have been achieved in this field in recent years.

Organized into 20 important chapters, this book incorporates the efforts and expertise of more than 30 outstanding experts, working alongside the editorial staff at Karger. The aim is to provide concise, practical, and comprehensive advanced knowledge. This will allow the physicians and researchers who are interested in the field of sleep medicine to catch up with current clinical knowledge and stay abreast of important developments that could be readily applied in daily practice and help meet the demands of our profession in this ever-evolving era.

Finally, I would like to thank all the authors for their great contributions, and the book series editor-in-chief Prof. Patrick Bradley and my mentor Prof. Michael Friedman for their tremendous dedication to the creation of this book. I hope all readers will enjoy this book and expand their knowledge. We definitely hope many will join us on the journey of sleep medicine.

Hsin-Ching Lin,
Kaohsiung City

Advances in Sleep-Related Breathing Disorders

What I knew, as a head and neck oncologic surgeon, could have been written on a "postage stamp," and I have memories of 2 patients presenting with sleep-related breathing difficulties, which on investigation were due to an underlying malignancy! During my early years of training in the 1970s, all snoring problems were "solved" by some form of nasal septal surgery with or without surgery to the oropharynx!

Sleep-related breathing disorders encompass a wide range of conditions, in which there is recurrent partial or complete cessation of breathing, and includes simple snoring, upper airway resistance syndrome, central sleep apnea-hypopnea syndrome, obesity-related sleep hypoventilation syndrome, and obstructive sleep apnea-hypopnea syndrome. Sleep-related breathing disorders, more so obstructive sleep apnea-hypopnea, can increase the risk of cardiac arrhythmias, pulmonary and systemic hypertension, myocardial infarction, type 2 diabetes mellitus, cerebrovascular accidents, impaired cognition, and road traffic accidents. The pathophysiology of sleep-related breathing disorders is complex, and the explanations usually involve several anatomical and other factors; obesity, increased neck circumference, craniofacial abnormalities and other variations (retrognathia, macroglossia, nasal obstructive pathologies), hypothyroidism, acromegaly, family history, excessive alcohol intake, or sedative ingestion. Many of the advances in the diagnosis and the formulation of modern treatment strategies have been developed through collaborative work in a multiprofessional and multidisciplinary environment.

While a number of "large textbooks" have been published on "sleep and breathing disorders" and "obstructive sleep apnea," this volume edited by Prof. Philip Lin (Hsin-Ching), Chairman of the Department of Otorhinolaryngology, and Director of Laryngology and Sleep Surgery, at Kaohsiung Chang Gung Memorial Hospital, Taiwan, has drawn upon his professional colleagues, friends and acquaintances from around the world to contribute chapters summarizing and updating knowledge on the new and innovative obstructive sleep apnea surgery and sleep medicine, as well as reviewing traditional knowledge on surgical and nonsurgical patient management.

This volume is a credit to the professionals and clinicians from many disciplines involved in all sleep-related breathing disorder clinics, who are to be congratulated on the progress that has been made over the past decades in unravelling this major health disorder. It has been a great pleasure for me to have invited and worked with Prof. Philip Hsin-Ching Lin in the maturation, development, and final publication of this volume.

Patrick J. Bradley, Nottingham

Advances in the Surgical Treatment of Obstructive Sleep Apnea

The future of sleep medicine and sleep surgery, like any other medical problem, is based on research, clinical studies, hard work, and commitment. After all, obstructive sleep apnea (OSA) is one of the most controversial, challenging, and ever-evolving specialties that otolaryngologists treat. The past few decades have seen a significant increase in research focusing on the pathophysiology, diagnosis, and surgical treatment of OSA. Great strides have been made in understanding how to treat OSA, and doctors must recognize that invasive and noninvasive treatments are crucial for the progress of the field. We must remember to resort to surgery only when patients have failed noninvasive treatment. A surgeon whose expertise includes both invasive and noninvasive options is in the ideal position to offer the most comprehensive treatment for OSA. These 2 treatment methods are not, however, mutually exclusive. Some patients may require an invasive treatment to enhance the efficacy of their long-term noninvasive treatment.

Sleep surgery is, and will likely always be, a suboptimal alternative to continuous positive airway pressure (CPAP). Arguments against sleep surgery are based on the low rate of surgical cures, but we must recognize that most diseases have low or nonexistent cure rates. Hypertension, diabetes, morbid obesity, and asthma are almost never cured, yet we do not deny palliative treatments to alleviate symptomatology and reduce the risk of long-term sequelae.

Even though surgery is less effective than CPAP, we should present all options and allow our patients to make informed decisions. Despite continuous improvements, CPAP compliance rates are not likely to reach 100%. A large percentage, and perhaps the majority, of patients with OSA have mild disease, and most of these patients defer testing and diagnosis because they are reluctant to accept CPAP therapy. Sleep surgery may be a beneficial and important treatment option for these patients.

This book represents the efforts of Philip Hsin-Ching Lin in all these aspects of sleep medicine and sleep surgery. Hsin-Ching's interest first began when he devoted his post-doc work to specific studies in sleep surgery. He was already a faculty member at Kaohsiung Chang Gung Memorial Hospital, but decided to leave his country and his family for a year in order to come to Chicago and study sleep medicine. Since then, he has made tremendous efforts to spend time with the giants of sleep medicine. In fact, he most recently travelled to Philadelphia to further his training in transoral robotic surgery. His hard work is clearly evident – he has over 80 peer-reviewed publications and has delivered close to 200 presentations worldwide. Among the 11 international awards he has received is the coveted International Guest Scholarships Award from the American College of Surgeons. Dr. Lin is a dynamic speaker and brilliant teacher, and is indeed well respected by all leaders in sleep medicine. The future of sleep medicine and sleep surgery is dependent on people who are dedicating their lives to clinical work, research, and teaching – and Philip Hsin-Ching Lin definitely embodies all these qualities.

Michael Friedman, Chicago, IL

Lin H-C (ed): Sleep-Related Breathing Disorders. Adv Otorhinolaryngol. Basel, Karger, 2017, vol 80, pp 1–6
DOI: 10.1159/000470486

A Succinct History of Sleep Medicine

Leh-Kiong Anne Huon • Christian Guilleminault
Stanford University Sleep Medicine Division, Redwood City, CA, USA

Abstract
Although sleep and sleep disorders have been studied for centuries, it was the development of sophisticated tools to monitor eye movements, brain waves, and muscle tone in the mid-20th century that led to modern sleep research. These tools allowed neuroscientists to distinguish between different "states" or "phases" of sleep, and to relate these findings to sleep disorders. This review chronicles the groundbreaking work of the pioneers in this field, and the impact their findings have had on patients today.

It was during the mid-20th century that our understanding of modern sleep was classified by modern research methods with the discovery of 3 states of alertness and a description of 2 states of sleep: NREM (non-rapid eye movement) and REM (rapid eye movement), not disregarding that sleep and its disturbed patterns had been recognized for centuries. Historically, pioneers such as Kleitman, Aserinsky, Dement, and Jouvet established the development of sleep medicine, indicating that sleep was not monolithic and that clear differences existed during sleep. Aserinsky and Kleitman [1] identified that eye movements were present during specific periods of sleep. Dement et al. [2] showed that brain waves were very different during sleep and that there were specific brain waves associated with these special eye movements. They also noted that periods of eye movement repeated at regular intervals of about every 100 min during nocturnal sleep. Jouvet and Michel [3] expanded the previous work and discovered that, in association with eye movements and brainwaves, there was an absence of muscle tone. In 1961 they categorized sleep into 2 different states: telencephalic (slow wave) sleep and rhombencephalic sleep (paradoxical sleep, known as REM sleep in English-language writing on the subject). This sent neuroscience toward a "new era." The use of monitoring of these 3 different variables during sleep (eye movements, brain waves, and muscle tone) allowed the investigation of sleep and its many alterations to sleep patterns. Dement founded the "Association for the Psycho-Physiology of Sleep," or APSS, which over time became the "Association of Professional Sleep Societies." Initially, during the early period, APSS membership consisted mainly of neuroscientists, psychiatrists interested in the potential relationship between dreaming and hallucinatory or delusory behavior, and neurologists interested in epilepsy and parasomnia or abnormal behavior during sleep. Narcolepsy, hypersomnia, the pharmacology of insomnia and sleep-related seizure disorders became the group's priorities. Parallel to this development, a group mostly interested in obesity wanted to understand the "hypersomnia" noted behind "massive obesity". Reading the "classic authors" and novels by Charles Dickens [4], Burwell et al. [5] wrote an article on "the Pickwickian syndrome," entitled "Extreme Obesity Associated with Alveolar Hypoventilation – A Pickwickian Syndrome."

The question was: what was the cause of the severe daytime somnolence observed in these obese individuals. The first recorded obese and sleepy "Pickwickian" patient presented for the investigation of recurring morning headaches and was observed to have respiratory pauses during sleep with recovery breathing accompanied by a loud snore. A polygraphic investigation was performed to explore the thoraco-abdominal breathing movements and the respiratory pauses while the subject was asleep; this patient was observed to associate his recovery breathing with a loud snore. A second patient also categorized as "Obese-Pickwickian" was again studied soon after and the findings were described as "periodic breathing with short suspension of breathing with the tongue falling back at onset of respiratory suspension, so there was no airflow despite increased movements of the thorax. Then the patient woke-up suddenly." These cases were reported by Gerardy, Herberg (Head of Internal Medicine, Heidelberg University Hospital), and Kuhn in 1962 [6]. The working hypothesis was that the daytime sleepiness was related to CO_2 retention. This hypothesis was further studied by the American researchers Drachman and Gumnit [7]. They also documented the associated changes in blood gases, again attributing the daytime sleepiness to CO_2 retention. This hypothesis was challenged by Wolfgang Kuhl, who changed his name to Kuhlo. Jung and Kuhlo [8] concluded that the sleepiness present in these patients was not due to increased CO_2, but proposed it to be related to a "central" component. Their patients underwent weight reduction and experienced a significant improvement in sleep and breathing functions, despite the fact that the syndrome did not disappear completely [8]. This work attracted the attention of Henry Gastaut from Marseille, who for many years had been an important member and prolific researcher of the European Neurophysiology Society, and Elio Lugaresi from Bologna. The respective universities of these researchers became central in the development of the understanding of abnormal breathing during sleep in obese subjects. Kuhlo et al. [9] had an obese-sleepy-Pickwickian that presented progressive and continuous decrease of consciousness leading to an emergency tracheostomy. Following such treatment, Kuhlo observed a great improvement of the daytime somnolence, with disappearance of the respiratory pauses during sleep. Kuhlo suggested treating these patients with a tracheostomy with cannula open during sleep. Before the procedure, the patient had not only nocturnal blood gases drawn at different times during the night, but also continuous measurement of their pulmonary artery pressure. After their surgery, the patients experienced a very rapid disappearance of daytime sleepiness. The seminal work and reports of Kuhl/Kuhlo had an important impact on his colleagues in Marseille and Bologna [10, 11], and the monitoring of breathing during sleep in Pickwickian patients became more sophisticated. Thoracic and abdominal bands were used, as was a thermocouple placed in a full-face mask. Both in Marseille and Bologna, changes in electroencephalography (EEG) patterns before and after the occurrence of an apnea episode were observed, and the term "respiratory-related arousal" was coined by the Marseille school [10]. The suggestion that the sleepiness seen in Pickwickian patients was related to this arousal pattern was made in 1966; this hypothesis was later confirmed by the results of tracheostomy. Work conducted in Strasbourg found that some patients only had partial apnea events that were called "hypopnea" by Kurtz (a neurophysiologist) and Lonsdorfer (a pulmonary specialist). These authors believed that the hypopneas indicated a transitional form of the Pickwickian syndrome either toward greater or lesser severity [12]. Following the report by Kuhl/Kuhlo that tracheostomy had significantly helped Pickwickian patients with apnea during sleep, patients were submitted to similar treatment approaches, and 5 of them were studied before and after tracheostomy with these invasive techniques during sleep. These studies occurred between 1970 and 1973, and other teams began performing similar investigations, particularly in Japan and the USA. The use of the term "Pickwickian syndrome" was questioned by the pulmonary specialist Douglas Carroll [13] in 1972:

> It is possible to identify a number of different syndromes which demonstrate either singly or in combination obesity, hypoventilation and hypersomnia. In the past many authors have referred to all these syndromes as Pickwickian syndrome. It appears that the waste basket of Pickwickian syndrome needs to be broken.

To this end, Lugaresi hoped to break up this "syndrome basket" and organized a meeting in Rimini, Italy, in 1972. A report was published in *Bulletin de Physio-Pathologie Respiratoire* (Clinical Respiratory Physiology), under the auspices of the French National Institute of Health and Medical Research (thereafter this journal merged with the *Scandinavian Respiratory Journal* to give birth to the *European Respiratory Journal*). What was ultimately reported was a broken hope, and the confusion remained unresolved. Nevertheless, the Rimini Symposium was a landmark as, for the first time, neurophysiologists and pulmonary specialists came together trying to unravel the different sleep-related breathing disorders that had been lumped together. The participants attempted to resolve the different clinical presentations, daytime consequences, and the underlying contributions. The presence of 3 types of polygraphic pattern was confirmed when studying breathing during sleep. It was recognized that esophageal manometry was the most accurate way at the time to define these 3 patterns, which were named "central," "obstructive," and "mixed" – the names still in use today. It also became clear that there were incomplete patterns that were called "hypopneas" that required further study. Simple recognition of the presence of a pattern was not diagnostic: the 3 types of pattern could be seen with a variable distribution in association with complaints of hypersomnia and obesity. Daytime testing for hypoventilation did not predict the presence or absence of hypoventilation during sleep. Carroll [13] proposed that the term "Pickwickian" should still be used, but that it should be subdivided based on the findings obtained after nocturnal polygraphy.

At the same congress, views on obstructive sleep apnea (OSA) as a pattern of Pickwickian Syndrome were also challenged. Guilleminault et al. [14] presented men of normal weight with typical OSA and complaints of insomnia, and also cases of narcoleptic-cataplectic patients with sleep apnea. These presentations challenged the supposition that OSA only occurred in obese, Pickwickian-type patients.

Also in 1973, Guilleminault et al. [14] reported pediatric cases of OSA. These authors described an "obstructive sleep apnea syndrome" (OSAS) [15] based on their studies of normal-weight subjects. These studies indicated that daytime sleepiness and insomnia could represent sleep-related complaints in such subjects. Subsequent investigations on the cardiovascular complications associated with OSA were undertaken. The typical brady-tachycardia associated with OSA and the cyclic variation of heart rate was described. The Stanford group performed a systematic investigation of the arrhythmias associated with OSA, indicating that asystole and atrial fibrillation were commonly seen in such patients [16]. Guilleminault et al. [17] also published a series of OSA children and their response to treatment with complete elimination of daytime hypertension, describing OSA syndrome in children. These advances were important and a new international meeting was planned by Guilleminault and Dement with the help of the Kroc Foundation in southern California in 1977. The subject was "Sleep Apnea Syndrome" [18], emphasizing the movement beyond "Pickwickian syndrome" and recognition of the condition's presence in subjects of normal weight. As a result of the meeting in 1977, the following year the different presentations were published and the existence of OSAS in adults was established [18]. In addition, the existence of cardiovascular comorbidity, specifically hypertension and cardiac arrhythmias, was confirmed. The role of the upper airway dilator muscles in the collapse of the upper airway became apparent. In short, this meeting presented the initial evidence that would become the subject of many research projects over the following 20 years.

The most important event thereafter, however, occurred in 1981 when Colin Sullivan and colleagues [19] reported the very beneficial results obtained with home continuous positive airway pressure treatment on OSA. Positive airway pressure had been used in neonatal intensive care, but Sullivan, using his mother's old vacuum cleaner, developed the equipment that eliminated the need to perform a tracheostomy. Moreover, by 1981, our understanding of OSAS in children was well advanced, and it was clear that the syndrome had a pervasive detrimental impact. It not only led to cardiovascular dysfunction, but there were clear effects on alertness, learning, memory, school success, growth, abnormal behavior suggestive of attention deficit and hyperactivity disorder, and mood disorders [20], including depression and parasomnias, such as enuresis, sleepwalking, and night-terrors, as shown by the Stanford group. The New York ORL-HNS department demonstrated that these symptoms could be helped by adenotonsillectomy [21].

If breathing during sleep had brought pulmonary specialists to the field of sleep, nobody believed in "sleep medicine." The term "sleep department" never emerged, but the sleep laboratory (probably one of the largest in Europe for a long time) was part of the Department of Neurology of the Montpellier Medical School in France. In some respects, the first International APSS congress in 1971 demonstrated the isolation of the few individuals who wanted to create a new research discipline, a primary goal of the APSS. In addition, the European neurologists realized that there was a complete absence of the concept of a clinical discipline, and that the APSS was essentially an American-based sleep research society, including PhD students with little clinical interest or expertise. At that time, none of the senior members of the APSS were interested in heading in that direction.

One individual who had a very different view was Christian Guilleminault. He was present at the 1970 APSS meeting and was clear about a new discipline that he called "sleep medicine." He opened a sleep medicine laboratory in La Salepetriere Hospital, against the desire of his chair and dean. He was financially supported by a 1-year grant from the "Societe Medicale des Hopitaux de Paris." In June 1971, he attended the first international meeting in Bruges and the future did not look promising: the same head was not supportive of his venture, despite the fact that the laboratory had been a success with more than 400 sleep-medicine patients monitored in a 2-bedroom EEG laboratory. At that time, Dement wanted to secure a large grant to investigate narcolepsy as a disorder of REM sleep, incorporating both animal and human research for a better understanding of REM sleep. Neurologists were not interested as it was mostly a low-paid research position based on a grant. Dement contacted Guilleminault in August 1971. Guilleminault was initially not very enthusiastic, considering the conditions and the focus on only narcolepsy and not on sleep medicine, including adults and children. As his situation deteriorated in France over the following months, Guilleminault agreed to go to Stanford, arriving in January 1972 with his goal of establishing "sleep medicine." Guilleminault and Dement had many discussions on the possibility of creating a clinical service where money would be obtained for services and tests. However, at the 1972 APSS meeting, the presentation of clinical results fell mostly on deaf ears. In order to gain a fee for the service, the field needed a well-defined test: nocturnal sleep recording and 24-h sleep recording for individuals with excessive daytime sleepiness as performed by Guilleminault were proposed, and Jerome Holland, a retired psychiatrist, coined the term "polysomnogram" in 1974. The only way to advance the medical arm of sleep research was to create an association of individuals interested in the clinical aspect of sleep. There were many people who would clearly participate, including Howard Roffward and Eliott Weitzman in New York, Ismet Karacan, Peter Hauri at Dartmouth, who was interested in insomnia, as was Milton Kramer, with a young PhD working in his department called Thomas Roth. All told, there were about 5 or 6 centers that would join, enough to create an official association. During 1975 there were frequent meetings at Stanford, and Dement was regularly travelling to the East Coast. Overall numbers were still low, but there were increasingly more places where clinical sleep-disorder patients were being seen. Finally, in the winter of 1975 in Chicago, the decision to create a new society called the Association of Sleep Disorder Centers (ASDC) was voted through, it would later change its name to the American Association of Sleep Medicine, and after 2000 to the American Academy of Sleep Medicine (AASM). In 1975, the APSS decided to create a journal entitled *Sleep* and nominated Guilleminault to take care of the articles and publication of the journal, and Dement to oversee the financing of the venture. Dement wanted to create a "sleep center" at the National Institutes for Health (NIH) and Guilleminault wanted to create "sleep medicine." An examination program to become specialists in sleep medicine began in 1975, it became an exam of the ASDC and then of the AASM, and was taken on by the Educational Council for General Medical Education in 2003. Dement had the creation of a "sleep center" at the NIH signed into law by US president Bill Clinton in 1993. An international effort was commenced in 1975 with the first International "Narcolepsy" symposium, which occurred in Montpellier and brought together sleep clinicians and scientists from all over the world [22].

Of the many sleep disorders, insomnia is the dominant condition. In the late 1970s, insomnia became recognized as a symptom more than a diagnosis, and treatment was directed to the underlying physical or psychological causes. A large number of different types of insomnia were recognized. The definition of "psycho-physiological insomnia" was presented in the first effort of the field to create a classification of sleep disorders, and in 1979 the first "Diagnostic Classification of Sleep and Arousal Disorders" was published [23]. One cause of sleep disturbance was the presence of leg movements during sleep. Termed "nocturnal myoclonus" by Symonds [24] in 1953, Lugaresi et al. [25] demonstrated that they were not related to a seizure disorder, and in 1975 Guilleminault et al. [26] demonstrated the repetitiveness and periodicity of these leg movements and their association with the complaint of poor sleep and insomnia when seen in isolation. Pursuing the work begun at Stanford, Coleman et al. [27] showed that these periodic movements could be seen in association with other sleep disorders, including narcolepsy and abnormal breathing during sleep, and the term "periodic leg movements" used in the 1975 article was subsequently changed to "periodic limb movements," recognizing that arms could also be involved. Lugaresi and collaborators renewed attention to "restless leg syndrome," initially described by Willis [28] and re-emphasized and studied in depth by Ekbom in 1944 [29]. In 1995, Walters et al. [30] published an improved definition of "restless leg syndrome," which led to the definition of diagnostic criteria to better recognize the problem in 2003 [31].

Sleep and sleep medicine were launched by the discoveries of the hypocretin-orexins [32, 33] and their role in alertness and in type 1 narcolepsy. Understanding of the many regions involved in the controls of sleep and wakefulness, as well as the interaction between sleep/wake and circadian rhythms advanced rapidly following these developments, establishing sleep medicine as a medical specialty.

References

1 Aserinsky E, Kleitman N: Regularly occurring periods of eye motility and concomitant phenomena during sleep. Science 1953;118:273–274.
2 Dement WC, Kleitman N: Cyclic variation in EEG during sleep and their relation to eye movements, body motility and dreaming. Clin Neurophysiol 1957; 9:673–690.
3 Jouvet M, Michel F: Sur les voies nerveuses responsables de l'activite rapide au cours du sommeil physiologique chez le chat (phase paradoxale). CR Soc Biol 1960;154:995–998.
4 Dickens C: The Posthumous Papers of the Pickwick Club. London, Chapman and Hall, 1837.
5 Burwell CS, Robin ED, Whaley RD, et al: Extreme obesity associated with alveolar hypoventilation – a Pickwickian syndrome. Am J Med 1956;21:811–818.
6 Gerardy W, Herberg D, Kuhn HM: Vergleichende Untersuchungen der Lungfunktion und der Elektroencephalogramms bei zwei Patienten mit Pickwickian Syndrom. Z Klin Med 1960; 156:362–380.
7 Drachman DB, Gumnit RJ: Periodic alteration of consciousness in the "Pickwickian" syndrome. Arch Neurol 1962; 6:471–477.
8 Jung R, Kuhlo W: Neurophysiology studies of abnormal night sleep in the Pickwickian syndrome; in Akert K, Bally C, Shade JP (eds): Progress in Brain Research: Sleep Mechanisms. Amsterdam, Elsevier, 1965, vol 18, pp 14–159.
9 Kuhlo W, Doll E, Franc MC: Erfolgreiche Behandlung eines Pickwick Syndrom durch eine dauertracheal Kanüle. Deutsch Med Wschr 1969;94:1286–1290.
10 Gastaut H, Tassinari CA, Duron B: Polygraphic study of the episodic diurnal and nocturnal manifestations of the Pickwickian syndrome. Brain Res 1966; 1:167–186.
11 Lugaresi E, Coccagna G, Petrella A, et al: Il disturbo de sonno e del respire nella syndrome di Pickwick. Sist Nerv 1968;1: 38–50.
12 Kurtz D, Meunier-Carus J, Bapst-Reiter J, Lonsdorfer J: Problemes nosologiques poses par certaines formes d'hypersomnie. Rev Electroencephalogr Neurophysiol Clin 1971;1:227–230.
13 Carroll D: Nosology of the Pickwickian syndrome. Bull Physiopathol Respir 1972;8:1241–1248.
14 Guilleminault C, Eldridge FL, Dement WC: Insomnia, narcolepsy and sleep apneas. Bull Physiopathol Respir 1972;8: 1127–1138.
15 Guilleminault C, Dement WC, Monod N: Syndrome "mort subite du nourrisson": apnees au cours du sommeil. Nouv Presse Med 1973;2:1355–1358.
16 Tilkian AG, Guilleminault C, Schroeder JS, et al: Sleep induced apnea syndrome: prevalence of cardiac arrhythmias and their reversal after tracheostomy. Am J Med 1977;63:348–358.
17 Guilleminault C, Eldridge F, Simmons F, Dement WC: Sleep apnea in eight children. Pediatrics 1976;58:23–30.
18 Guilleminault C, Dement WC (eds): Sleep Apnea Syndrome. New York, AR Liss, 1978, pp 1–372.
19 Sullivan CE, Berthon-Jones M, Issac FC, et al: Reversal of obstructive sleep apnea by continuous positive airway pressure applied through the nares. Lancet 1981; 1:862–865.
20 Guilleminault C, Korobkin R, Winkle R: A review of 50 children with obstructive sleep apnea syndrome. Lung 1981;159: 275–287.
21 Kravath R, Pollak C, Borowiecki B: Hypoventilation during sleep in children who have lymphoid airway obstruction treated by a nasopharyngeal tube and T and A. Pediatrics 1977;59:865–871.
22 Guilleminault C, Dement WC, Passouant P (eds): Narcolepsy: Advances in Sleep Research. New York, Spectrum, 1976, vol 3, pp 1–684.
23 Association of Sleep Disorder Centers, Sleep Disorders Classification Committee: Diagnostic classification of sleep and arousal disorders. Sleep 1979;2:1–137.
24 Symonds CP: Nocturnal myoclonus. J Neurol Neurosurg Psychiatr 1953;16: 166.
25 Lugaresi E, Coccagna G, Berti Cerroni G, et al: Restless leg syndrome and nocturnal myoclonus; in Gastaut H, Lugaresi E, Berti Cerroni G (eds): The Abnormalities of Sleep in Man. Bologna, Aulo Gaggi, 1968, pp 285–294.
26 Guilleminault C, Raynal D, Weitzman ED, et al: Sleep related periodic myoclonus in patients complaining of insomnia. Trans Am Neurol Assoc 1975;100: 19–22.
27 Colemann RM, Pollack CP, Weitzman ED: Periodic movements in sleep (nocturnal myoclonus): relation to sleep disorders. Ann Neurol 1980;8:416–421.
28 Willis T: The London Practice of Physicks. London, Bassett and Crooke, 1685.
29 Ekbom KA: Asthenia crurum paraesthetica (irritable legs). Acta Med Scand 1944;118:197–209.
30 Walters AS, Aldrich MA, Allen RP, et al: Toward a better definition of the restless leg syndrome. Mov Disord 1995;10:634–642.
31 Allen RP, Picchietti D, Hening WA, et al: Restless legs syndrome: diagnostic criteria, special considerations, and epidemiology. A report from the Restless Leg Syndrome Diagnosis and Epidemiology Workshop at the National Institutes of Health. Sleep Med 2003;4:101–119.
32 De Lecea L, Kilduff TS, Peyron C, et al: The hypocretins: hypothalamus specific peptides with neuro-excitory activity. Proc Natl Acad Sci USA 1998;95:322–327.
33 Sakurai T, Amemiya A, Ishii M, et al: Orexins and orexin receptors: a family of hypothalamic neuropeptides and G protein-coupled receptors that regulate feeding behavior. Cell 1998;92:573–585.

Prof. Christian Guilleminault, MD, Biol D
Stanford University Sleep Medicine Division
450 Broadway Street, Pavilion B, 2nd Floor, MC 5730
Redwood City, CA 94063 (USA)
E-Mail cguil@stanford.edu

Lin H-C (ed): Sleep-Related Breathing Disorders. Adv Otorhinolaryngol. Basel, Karger, 2017, vol 80, pp 7–16
DOI: 10.1159/000470522

Advanced Concepts in the Pathophysiology of Obstructive Sleep Apnea

David P. White

Brigham and Women's Hospital, Harvard Medical School, Boston, MA, USA

Abstract

The primary pathological event in the disorder obstructive sleep apnea (OSA) is the partial or complete closure of the pharyngeal airway during sleep in an individual with a widely patent airway during wakefulness. This yields an apnea or hypopnea with resulting hypoxia and hypercapnia, and most often requires an arousal to terminate the event. These events occur in a repetitive manner during sleep, yielding intermittent hypoxia and sleep fragmentation with their associated adverse effects on the health and quality of life of the afflicted individual. Here, we focus on the events leading up to these apneas and hypopneas, primarily addressing pharyngeal anatomy, upper airway muscle control, the respiratory arousal threshold, and ventilatory control instability in OSA pathophysiology.

Pharyngeal Anatomy

To develop sleep apnea, there must be some anatomically increased predisposition to pharyngeal collapse. Obstructive apnea will not develop in an individual with a large pharyngeal airway in the absence of substantial pharyngeal dilator muscle activity. To this end, there are abundant imaging studies, generally conducted during wakefulness, demonstrating an anatomical smaller upper airway in patients with obstructive sleep apnea (OSA) when compared to normal controls [1–3]. This is generally the product of either a small bony airway enclosure (mandible, maxilla, and cervical spine) or increased soft tissue in a normal-sized enclosure [4]. Thus, airway size is an interaction between the bony skeleton and the enclosed soft tissue.

Physiological studies, generally conducted during sleep, have also indicated increased airway collapsibility in patients with OSA compared to controls. The most commonly used metric is the critical closing pressure or Pcrt [5, 6]. This is determined by progressively dropping the continuous positive airway pressure (CPAP) to a level at which the airway closes. In normal controls, this is generally a pressure less than negative 5–6 cm H_2O. Patients with OSA have Pcrt levels higher (less negative) than negative 5 cm H_2O and often have a positive closing pressure.

Although the metrics described above were designed to measure anatomy, they can certainly

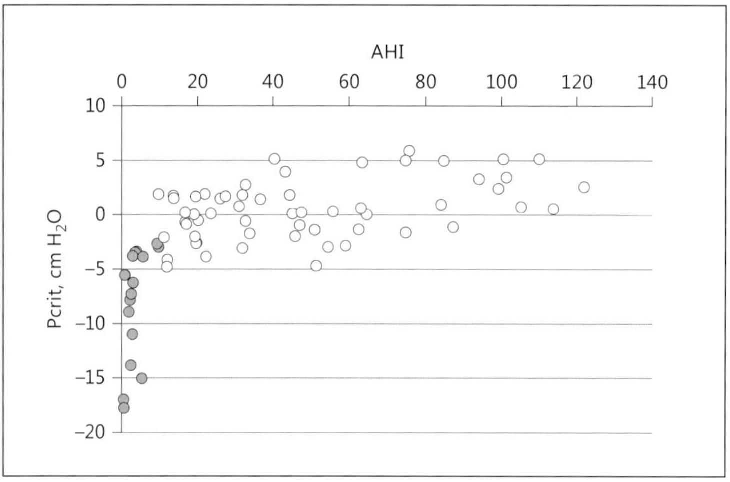

Fig. 1. Critical closing pressure of the upper airway (Pcrit) versus AHI. Gray circles represent individuals with an AHI <10. Reprinted with permission from Eckert et al. [11].

be influenced by physiologic events as well. The 2 such events clearly demonstrated to influence airway collapsibility are lung volume and fluid redistribution. Increases in lung volume, produced in a number of ways, can lead to decreased pharyngeal airflow resistance during sleep and lower CPAP requirements to eliminate apnea [7, 8]. Opposite effects are seen with a decreasing lung volume. This is most likely the product of increased longitudinal traction on the trachea and upper airway as the lung volume rises, yielding a stiffer airway. The decrements in lung volume reported during sleep may therefore contribute to increased collapsibility. A substantial literature is also evolving, suggesting that fluid redistribution in the supine posture during sleep may lead to an increased fluid volume in the neck, yielding a small airway that is more susceptible to collapse [9]. Thus, pure anatomy and these physiologic events are both important.

The importance of this abnormal anatomy in the pathophysiology of OSA should not be underestimated. However, anatomy is far from the entire explanation for the development of airway collapse during sleep. Any assessment of airway anatomy, be it imaging awake or Pcrt measured during sleep, shows considerable overlap between patients with OSA and normal subjects [10]. Recent assessment of Pcrt by Eckert at al. [11] revealed that virtually all individuals with a Pcrt >–2 cm H_2O had OSA, while a Pcrt <–5 cm H_2O was only found in normal subjects (Fig. 1). However, a Pcrt between –5 and –2 cm H_2O was found in some OSA patients and some normal controls. In addition, as can be seen in Figure 1, at all Pcrt levels there was considerable variability in apnea severity. At the same Pcrt of 0 cm H_2O, some patients had an apnea-hypopnea index (AHI) of about 15–20, while others had an AHI >110. Thus, anatomy is far from the whole story of OSA pathophysiology.

Upper Airway Motor Control

The importance of pharyngeal muscle control in OSA pathophysiology is abundantly clear from the state-related nature of the disorder [12]. Even patients with the most severely compromised upper airway anatomy have little to no problem maintaining airway patency during wakefulness. It is only with sleep onset that problems begin. Although many things may change when an individual falls asleep, the principle influence on the

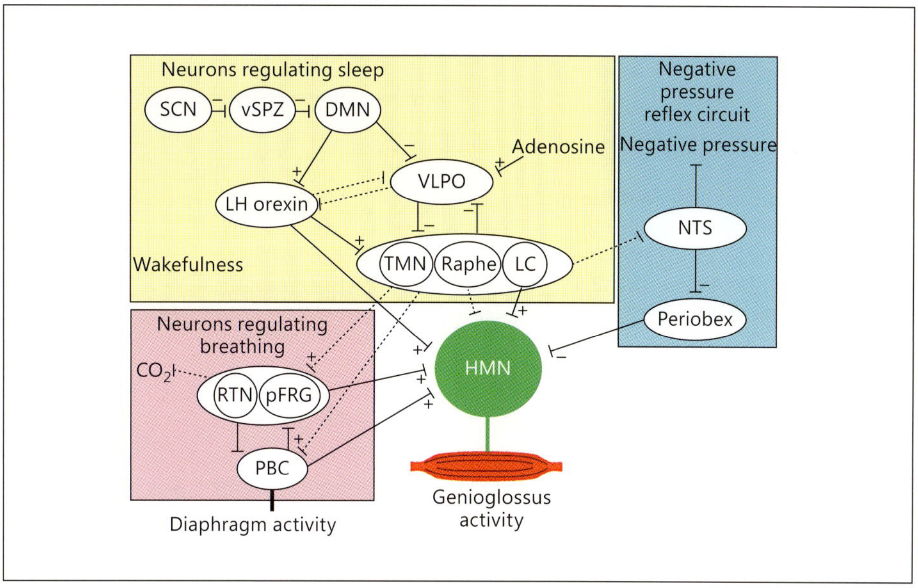

Fig. 2. Proposed control of hypoglossal motor output to the genioglossal muscle. The hypoglossal motor nucleus (HMN) receives input from neurons involved in negative pressure reflex circuits, and those controlling sleep and breathing. Dotted lines represent proposed pathways with the exception of the raphe, where connections to the HMN are well documented but not believed to contribute to state-related changes in HMN activity. SCN, suprachiasmatic nucleus; vSPZ, ventral subparaventricular zone; DMN, dorsomedial hypothalamic nucleus; VLPO, ventrolateral preoptic nucleus; LH, orexin lateral hypothalamic orexinergic neurons; LC, locus coeruleus; TMN, tuberomammillary nucleus; pFRG, parafacial respiratory group; RTN, retrotrapezoid nucleus; PBC, pre-Bötzinger complex; NTS, nucleus of the solitary tract. Some arrows have been omitted for clarity. Reprinted with permission from Jordan and White [12].

patency of the pharyngeal airway is a decrement in the activity of dilator muscles.

There are numerous pharyngeal dilator muscles, which fall into 4 main groups [13]: (a) extrinsic tongue muscles (e.g., genioglossus, GG), (b) muscles controlling palatal stiffness and position (e.g., tensor palatini and levator palatini), (c) pharyngeal constrictors, and (d) muscles controlling hyoid bone position (e.g., geniohyoid and sternohyoid). These muscles are quite different in their origins and insertions, patterns of activation (phasic with respiration or tonic), and the influence of sleep on their activity. A full analysis of these various muscles is beyond the scope of this discussion (see Jordan and White [12] for full details). Thus, the focus will be on the GG, an important pharyngeal dilator muscle that has been extensively studied.

Wakefulness

The activity of the GG muscle, when quantified using multiunit recordings, is primarily controlled by 3 inputs: (a) airway negative pressure [14], (b) respiratory drive (and thereby hypoxia and hypercapnia) [15], and (c) state-dependent neuromodulators [16] (Fig. 2). Abundant evidence indicates that negative pressure in the airway leads to an increase in genioglossal electromyographic (GGEMG) activity [14]. This is believed to be the primary mechanism by which pharyngeal dilator muscles protect the airway when it is challenged by collapsing forces. The same basic principle applies to respiratory drive. When an individual is made hypoxic or hypercapnic, which increase respiratory drive, genio-

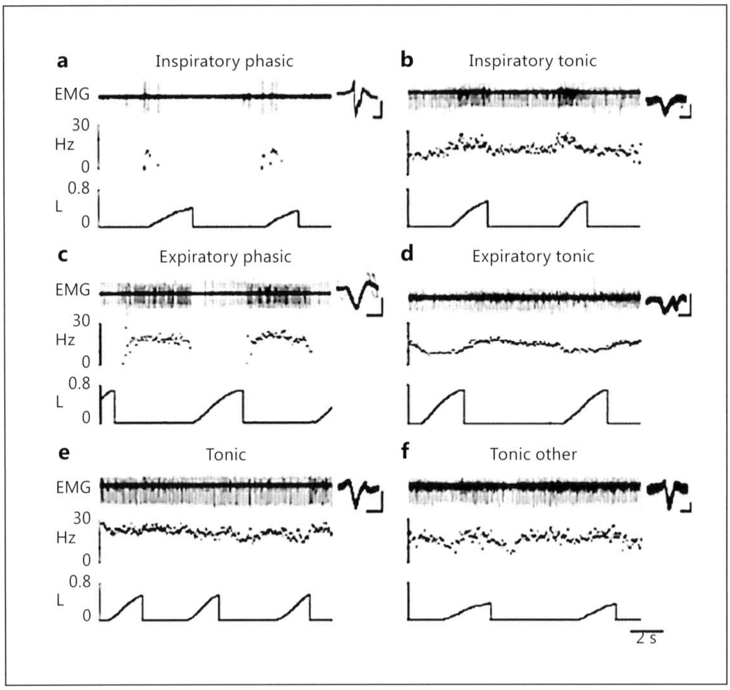

Fig. 3. GG motor-unit types. **a–f** Typical examples of the 6 types of motor-unit discharge recorded from the GG during quiet breathing. The raw electromyogram (EMG), instantaneous discharge frequency plot for 1 unit derived from the raw signal and inspired volume are shown for each type of unit. Action potentials from the single units for 2 breaths are superimposed to the right of each panel. Vertical calibration bars for the EMG are 500 μV for **a**, **b**, and **f**, and 200 μV for **c–e**, and the horizontal calibration bar is 2 ms. Recordings are from 3 different subjects. Reprinted with permission from Saboisky et al. [19].

glossal activity increases [15]. This likely results from both direct inputs from respiratory pattern-generating neurons (Fig. 2) and from the increased negative pressure that develops in the airway as respiratory effort increases. This again protects the airway patency as the respiratory drive increases. Finally, various excitatory neurotransmitters, such as norepinephrine, steadily activate the hypoglossal motoneurons during wakefulness, thereby increasing genioglossal activity [17].

It should be recognized, however, that multiunit recording of the GGEMG underestimates the complexity of this muscle. Recent work using single motor unit recording techniques [18, 19] indicates that there are at least 6 different firing patterns for GG motor units, ranging from inspiratory phasic, to expiratory phasic, to purely tonic activity (Fig. 3). This speaks to the precision needed to control tongue position, shape, and stiffness. Keeping the airway open is but one of the functions of the genioglossal muscle. These single motor unit recordings will not be addressed further here, but it must be kept in mind that multiunit recording provides only a crude estimate of what is happening in this complex muscle.

Most evidence would suggest that during wakefulness the activity of the GG and other upper airway dilator muscles is increased in OSA patients compared to normal controls [20]. This increased activity is believed to be largely driven by the negative pressure reflex in response to the compromised pharyngeal anatomy present in the OSA patient, as described above [21]. Thus, the airway dilator muscles compensate for the deficient anatomy and maintain pharyngeal patency quite comfortably during wakefulness. However, this interpretation has been challenged, with some investigators believing the increased multiunit GGEMG activity of the OSA patient to be a product of a myopathy rather than increased neural input [22]. Most evidence, however, supports the concept that neural drive to the GGEMG is increased in OSA patients in response to a poor anatomy.

Sleep

With sleep onset, there are changes in all components of genioglossal regulation. First, due to the withdrawal of excitatory neurotransmitters (norepinephrine and serotonin) at the hypoglossal motor nucleus, there is a general reduction in the activity of the muscle [16]. Second, substantial evidence indicates that there is a reduction in the negative pressure responsiveness of this muscle during non-rapid eye movement (NREM) sleep compared to wakefulness in normal subjects [23]. Fewer data are available on this topic during REM sleep [24]. Finally, CO_2 responsiveness of the GG muscle during sleep has been minimally studied, but the data available suggest a decreased slope of the GGEMG/PCO_2 response [25]. Thus, all drives to the GG seem to be reduced during NREM sleep. However, this should not suggest that the muscle is unresponsive during sleep, as this is not the case. The muscle can clearly respond during sleep to all stimuli, although in many subjects less briskly and possibly more slowly than during wakefulness [23, 26].

All of the information provided above would suggest that genioglossal muscle activity should fall during sleep and should probably fall more in patients with OSA, who have increased waking activity, than normal controls. There is a body of work that supports this concept, demonstrating substantial decrements in GGEMG activity in the initial breaths following sleep onset [27, 28]. In addition, data acquired over the entire night in OSA patients indicates lower levels of GGEMG activity in association with apneas and hypopneas than during periods of stable breathing [29]. Thus, decrements in pharyngeal dilator muscle activity are almost certainly an important contributor to upper airway collapse.

Despite the findings described above, the GG muscle is still responsive to standard stimuli during sleep and often quickly recovers from the initial sleep-induced decrements to reach levels as high as or higher than those observed during wakefulness [30]. As a result, OSA patients may demonstrate apneas and hypopneas at times when there is considerable genioglossal activation. What accounts for the increased pharyngeal collapsibility at these times is not completely clear, but could result from changes in lung volume, decrements in the activity of other pharyngeal dilator muscles, or other unmeasured events, such as changes in jaw position. Further research is needed to clarify the cause of such events.

Finally, it must be recognized that there is considerable variability between individuals in the ability of the upper airway to respond to a collapsing challenge [11, 31]. In some individuals, when the upper airway is challenged with progressively negative collapsing pressures, the airway quickly closes and does not reopen until the individual awakens. In others, challenges with similar or greater collapsing pressures will lead to a brisk muscle response, which actively protects airway patency. Thus, some individuals with poor airway anatomy are able to maintain an open airway during sleep, while others with a less severe anatomic problem will readily allow the airway to close at sleep onset. Thus, anatomy and pharyngeal muscle responsiveness are separate traits that together importantly dictate who does and does not develop OSA.

Respiratory Arousal Threshold

In most individuals, when the respiratory system is adequately driven, arousal from sleep will occur in response to this stimulus. It is generally accepted that it is respiratory drive that ultimately awakens the individual and not the source of the elevated drive, such as hypoxia and hypercapnia [32, 33]. In addition, there is wide variability in this

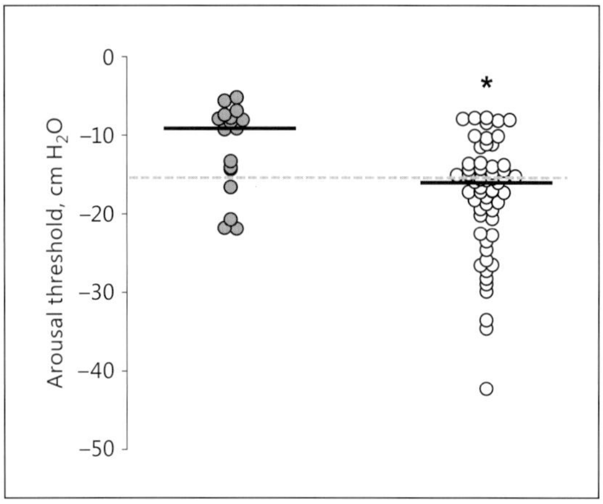

Fig. 4. Phenotypic scatter plot of the respiratory arousal threshold. Notice the large degree of between-subject variability within patients with OSA and control subjects, and considerable between-group overlap. Horizontal black lines represent median values. Values above the dashed grey lines indicate a low arousal threshold. The values for the 2 groups are statistically different, * $p < 0.05$. Reprinted with permission from Eckert et al. [11].

respiratory arousal threshold from 1 individual to the next [34], both in normal controls and in patients with OSA (Fig. 4). Thus, some individuals wake up in response to minor changes in respiratory drive while others require a very elevated drive before arousal will occur.

There are at least 2 ways in which this respiratory arousal threshold can influence OSA pathophysiology. First, based on the discussion above, if a given individual has a small, collapsible pharyngeal airway, but also responsive upper airway muscles while asleep, that individual must be able to stay asleep long enough for those pharyngeal dilator muscles to be recruited to stabilize the airway [35]. Thus, the individual must fall asleep and experience a decrease in pharyngeal patency yielding some decrement in minute ventilation. This will lead to a slow increase in respiratory drive (due to rising PCO_2) with increasing airway negative pressure. Both the slowly increasing drive and falling airway pressure will activate pharyngeal dilator muscles. If the individual can stay asleep long enough for these muscles to become adequately active, pharyngeal patency will be restored and stable sleep and breathing established. On the other hand, if the individual awakens every time there is a small decrease in minute ventilation with a minor increase in respiratory drive, stable sleep with a patent airway can never be achieved. As a result, the individual will cycle between sleep and wake, and hypo- and hyperventilation. Thus, in an individual with a poor anatomy and a good upper airway muscle response during sleep, a high respiratory arousal threshold is desirable if stable sleep and breathing is to be achieved. On the other hand, in an individual with a poor upper airway response, a high respiratory arousal threshold will only lead to longer apneas and hypopneas with greater hypoxia and hypercapnia.

The second way in which the respiratory arousal threshold can influence apnea pathophysiology is through the potentially destabilizing influence of arousal from sleep on ventilatory control. Some believe that a robust arousal from sleep in response to an apnea or hypopnea may lead to excessive hyperventilation with subsequent hypocapnia [36]. This low PCO_2 can lead to reduced ventilator drive and falling pharyngeal dilator muscle activity, which may contribute to the development of the next apnea or hypopnea. Thus, 1 sleep disordered breathing event propagates the next through arousal-induced hyperventilation. If arousal can be avoided and airway patency reestablished without arousal, the hyperventilation might well be minimized and the cycling ventilation reduced or eliminated. However, not all data support this concept, with some studies reporting little relationship between the presence or absence of arousal from sleep following an apnea/hypopnea and the subsequent predisposition to the next event [37]. Thus, the potentially destabilizing role of arousal from sleep on ventilatory control remains controversial.

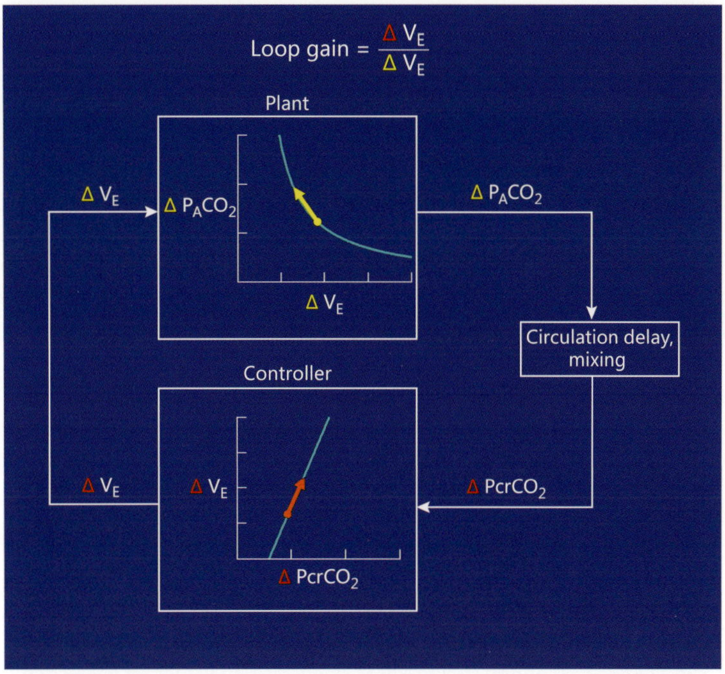

Fig. 5. Demonstration of the feedback loop for the respiratory system that drives loop gain. There are 2 main components: the plant and the controller. The plant gain consists primarily of the lungs and respiratory muscles and its gain quantified as the change in arterial PCO_2 that will occur with a change in ventilation (ΔV_E, the disturbance). Thus, in the case of OSA, the plant dictates how much PCO_2 will rise in response to an apnea or hypopnea. The controller gain is dictated by the chemoreceptor for arterial PCO_2 (carotid body and retrotrapezoid nucleus) generally referred to as the hypercapnic ventilatory response. Thus, a change in arterial PCO_2 due to an apnea/hypopnea (the disturbance), following a circulatory delay, will lead to a response (ΔV_E) based on the slope of the hypercapnic ventilatory response. Loop gain is defined as $\Delta V_E / \Delta V_E$.

Loop Gain (Ventilatory Control Instability)

The role of ventilatory control instability in the pathogenesis of OSA has been debated for years, with some believing that an overshoot in ventilation following an apnea or hypopnea leads to a subsequent depression of respiratory drive that contributes to the next obstructive event [38]. Thus, cycling ventilation is perpetuated. Others believe that such respiratory cycling is primarily due to upper airway obstruction with little role for variability in ventilatory control in this process.

More recently the concept of loop gain has emerged [39], with loop gain being an engineering term describing the gain of any physiologic system controlled by feedback loops (Fig. 5). Loop gain is a unitless term generally defined as the ratio of ventilatory response/ventilatory disturbance, with a higher loop gain indicating more unstable respiratory control. Thus, a large response to a small respiratory disturbance indicates a high loop gain and unstable ventilatory control.

There is not a lot of evidence that patients with OSA have a systematically higher loop gain than normal controls [11], although severe OSA patients may have a higher loop gain than patients with less severe OSA [39]. That being said, high loop gain does predispose a patient to the development of OSA if the other traits described above are also in alignment. This primarily means that the patient, in addition to a high loop gain, has a substantially collapsible pharyngeal airway. Eckert et al. [11] reported that 50% of patients with a moderately collapsible pharyngeal airway had an elevated loop gain, suggesting an important role for this trait in apnea pathogenesis in this particular group of patients.

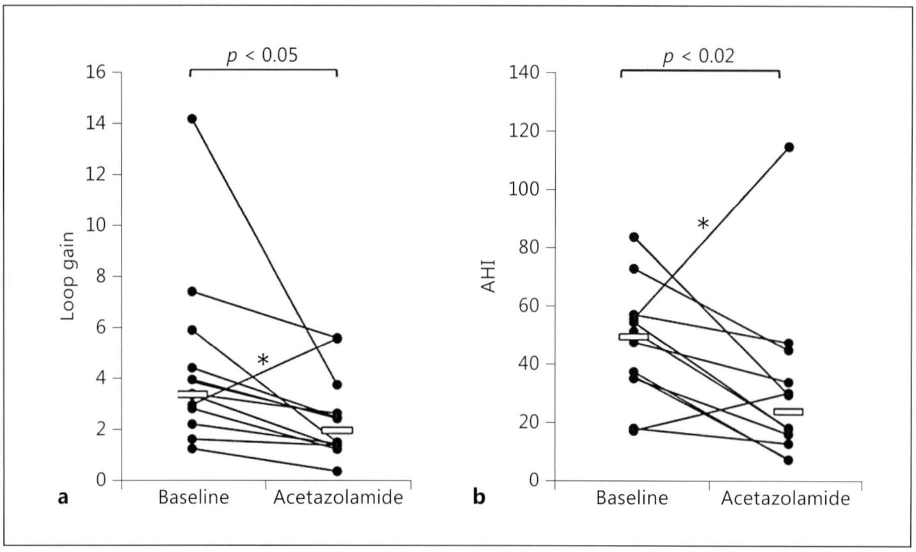

Fig. 6. Individual effects of acetazolamide on loop gain and the severity of sleep-disordered breathing. **a** Individual data comparing steady-stage loop gain during baseline and acetazolamide conditions. Note that the steady-state loop gain was reduced in all but 1 subject. **b** Acetazolamide significantly reduced the median AHI from 49.6 (35.5–56.6) to 24.1 (12.9–42.3) during supine NREM sleep. Median values for each condition are indicated by the solid white bars. Note that 1 patient whose loop gain doubled with acetazolamide also demonstrated a similar rise in AHI (asterisk). Reprinted with permission from Edwards et al. [40].

To truly grasp the importance of loop gain as a cause of OSA, this trait must be manipulated and the effect of this maneuver on OSA severity assessed. Several studies have now accomplished this using both oxygen and acetazolamide to reduce loop gain. Edwards et al. [40] reported a 47% overall reduction in AHI following 1 week of acetazolamide administration (Fig. 6). Similar observations have been made with oxygen, with a stronger response being observed in patients with an initially high loop gain [41]. These studies strongly suggest that a high loop gain is an important pathophysiologic trait in the development of OSA in many patients.

Conclusion

Current evidence suggests that the cause of OSA is multifactorial and that this pathophysiology varies substantially from 1 patient to the next. All OSA patients must have some anatomical predisposition to pharyngeal collapse, but anatomy alone explains little of the variability in apnea severity between patients. Thus, other physiologic traits are fundamentally important in the development of this disorder in many patients. Although measurement of these traits as a guide to apnea treatment is not commonly utilized at this time, this will likely change over time as methods to measure the traits are simplified and novel ways to manipulate them identified.

References

1 Haponik EF, Smith PL, Bohlman ME, Allen RP, Goldman SM, Bleecker ER: Computerized tomography in obstructive sleep apnea: correlation of airway size with physiology during sleep and wakefulness. Am Rev Respir Dis 1983; 127:221–226.

2 Schwab RJ, Gupta KB, Gefter WB, Metzger LJ, Hoffman EA, Pack AI: Upper airway and soft tissue anatomy in normal subjects and patients with sleep-disordered breathing: significance of the lateral pharyngeal walls. Am J Respir Crit Care Med 1995;152:1673–1689.

3 Schwab RJ, Pasirstein M, Pierson R, Mackley A, Hachadoorian R, Arens R, Maislin G, Pack AI: Identification of upper airway anatomic risk factors for obstructive sleep apnea with volumetric magnetic resonance imaging. Am J Respir Crit Care Med 2003;168:522–530.

4 Watanabe T, Isono S, Tanaka A, Tanzawa H, Nishino T: Contribution of body habitus and craniofacial characteristics to segmental closing pressures of the passive pharynx in patients with sleep-disordered breathing. Am J Respir Crit Care Med 2002 15;165:260–265.

5 Schwartz AR, Smith PL, Wise RA, Gold AR, Permutt S: Induction of upper airway occlusion in sleeping individuals with subatmospheric nasal pressure. J Appl Physiol (1985)1988;64:535–542.

6 Kirkness JP, Peterson LA, Squier SB, McGinley BM, Schneider H, Meyer A, Schwartz AR, Smith PL, Patil SP: Performance characteristics of upper airway critical collapsing pressure measurements during sleep. Sleep 2011;34:459–467.

7 Heinzer RC, Stanchina ML, Malhotra A, Fogel RB, Patel SR, Jordan AS, Schory K, White DP: Lung volume and continuous positive airway pressure requirements in obstructive sleep apnea. Am J Respir Crit Care Med 2005;172:114–117.

8 Heinzer RC, Stanchina ML, Malhotra A, Jordan AS, Patel SR, Lo YL, Wellman A, Schory K, Dover L, White DP: Effect of increased lung volume on sleep disordered breathing in patients with sleep apnoea. Thorax 2006;61:435–439.

9 Redolfi S, Yumino D, Ruttanaumpawan P, Yau B, Su MC, Lam J, Bradley TD: Relationship between overnight rostral fluid shift and obstructive sleep apnea in nonobese men. Am J Respir Crit Care Med 2009;179:241–246.

10 Sforza E, Petiau C, Weiss T, Thibault A, Krieger J: Pharyngeal critical pressure in patients with obstructive sleep apnea syndrome: clinical implications. Am J Respir Crit Care Med 1999;159:149–157.

11 Eckert DJ, White DP, Jordan AS, Malhotra A, Wellman A: Defining phenotypic causes of obstructive sleep apnea: identification of novel therapeutic targets. Am J Respir Crit Care Med 2013; 188:996–1004.

12 Jordan AS, White DP: Pharyngeal motor control and the pathogenesis of obstructive sleep apnea. Respir Physiol Neurobiol 2008;160:1–7.

13 White DP, Younes MK: Obstructive sleep apnea. Compr Physiol 2012;2: 2541–2594.

14 Akahoshi T, White DP, Edwards JK, Beauregard J, Shea SA: Phasic mechanoreceptor stimuli can induce phasic activation of upper airway muscles in humans. J Physiol 2001;531:677–691.

15 Stanchina ML, Malhotra A, Fogel RB, Ayas N, Edwards JK, Schory K, White DP: Genioglossus muscle responsiveness to chemical and mechanical stimuli during non-rapid eye movement sleep. Am J Respir Crit Care Med 2002;165:945–949.

16 Lo YL, Jordan AS, Malhotra A, Wellman A, Heinzer RA, Eikermann M, Schory K, Dover L, White DP: Influence of wakefulness on pharyngeal airway muscle activity. Thorax 2007;62:799–805.

17 Horner RL: Neuromodulation of hypoglossal motoneurons during sleep. Respir Physiol Neurobiol 2008;164:179–196.

18 Saboisky JP, Butler JE, Fogel RB, Taylor JL, Trinder JA, White DP, Gandevia SC: Tonic and phasic respiratory drives to human genioglossus motoneurons during breathing. J Neurophysiol 2006;95: 2213–2221.

19 Saboisky JP, Butler JE, McKenzie DK, Gorman RB, Trinder JA, White DP, Gandevia SC: Neural drive to human genioglossus in obstructive sleep apnoea. J Physiol 2007;585:135–146.

20 Mezzanotte WS, Tangel DJ, White DP: Waking genioglossal electromyogram in sleep apnea patients versus normal controls (a neuromuscular compensatory mechanism). J Clin Invest 1992;89: 1571–1579.

21 Fogel RB, Malhotra A, Pillar G, Edwards JK, Beauregard J, Shea SA, White DP: Genioglossal activation in patients with obstructive sleep apnea versus control subjects: mechanisms of muscle control. Am J Respir Crit Care Med 2001;164: 2025–2030.

22 Saboisky JP, Stashuk DW, Hamilton-Wright A, Carusona AL, Campana LM, Trinder J, Eckert DJ, Jordan AS, McSharry DG, White DP, Nandedkar S, David WS, Malhotra A: Neurogenic changes in the upper airway of patients with obstructive sleep apnea. Am J Respir Crit Care Med 2012;185:322–329.

23 Wheatley JR, Mezzanotte WS, Tangel DJ, White DP: Influence of sleep on genioglossus muscle activation by negative pressure in normal men. Am Rev Respir Dis 1993;148:597–605.

24 Shea SA, Edwards JK, White DP: Effect of wake-sleep transitions and rapid eye movement sleep on pharyngeal muscle response to negative pressure in humans. J Physiol 1999;520:897–908.

25 Lo YL, Jordan AS, Malhotra A, Wellman A, Heinzer RC, Schory K, Dover L, Fogel RB, White DP: Genioglossal muscle response to CO_2 stimulation during NREM sleep. Sleep 2006;29:470–477.

26 Wiegand L, Zwillich CW, White DP: Collapsibility of the human upper airway during normal sleep. J Appl Physiol (1985) 1989;66:1800–1808.

27 Mezzanotte WS, Tangel DJ, White DP: Influence of sleep onset on upper-airway muscle activity in apnea patients versus normal controls. Am J Respir Crit Care Med 1996;153:1880–1887.

28 Fogel RB, Trinder J, White DP, Malhotra A, Raneri J, Schory K, Kleverlaan D, Pierce RJ: The effect of sleep onset on upper airway muscle activity in patients with sleep apnoea versus controls. J Physiol 2005;564:549–562.

29 Jordan AS, White DP, Lo YL, Wellman A, Eckert DJ, Yim-Yeh S, Eikermann M, Smith SA, Stevenson KE, Malhotra A: Airway dilator muscle activity and lung volume during stable breathing in obstructive sleep apnea. Sleep 2009;32: 361–368.

30 Basner RC, Ringler J, Schwartzstein RM, Weinberger SE, Weiss JW: Phasic electromyographic activity of the genioglossus increases in normals during slow-wave sleep. Respir Physiol 1991;83: 189–200.

31 Jordan AS, Wellman A, Heinzer RC, Lo YL, Schory K, Dover L, Gautam S, Malhotra A, White DP: Mechanisms used to restore ventilation after partial upper airway collapse during sleep in humans. Thorax 2007;62:861–867.
32 Gleeson K, Zwillich CW, White DP: The influence of increasing ventilatory effort on arousal from sleep. Am Rev Respir Dis 1990;142:295–300.
33 Berry RB, Gleeson K: Respiratory arousal from sleep: mechanisms and significance. Sleep 1997;20:654–675.
34 Eckert DJ, Younes MK: Arousal from sleep: implications for obstructive sleep apnea pathogenesis and treatment. J Appl Physiol (1985) 2014;116:302–313.
35 Younes M, Loewen AH, Ostrowski M, Laprairie J, Maturino F, Hanly PJ: Genioglossus activity available via non-arousal mechanisms vs. that required for opening the airway in obstructive apnea patients. J Appl Physiol (1985) 2012;112:249–258.
36 Younes M: Role of arousals in the pathogenesis of obstructive sleep apnea. Am J Respir Crit Care Med 2004;169:623–633.
37 Jordan AS, Cori JM, Dawson A, Nicholas CL, O'Donoghue FJ, Catcheside PG, Eckert DJ, McEvoy RD, Trinder J: Arousal from sleep does not lead to reduced dilator muscle activity or elevated upper airway resistance on return to sleep in healthy individuals. Sleep 2015;38:53–59.
38 Onal E, Lopata M: Periodic breathing and the pathogenesis of occlusive sleep apneas. Am Rev Respir Dis 1982;126: 676–680.
39 Younes M, Ostrowski M, Thompson W, Leslie C, Shewchuk W: Chemical control stability in patients with obstructive sleep apnea. Am J Respir Crit Care Med 2001;163:1181–1190.
40 Edwards BA, Sands SA, Eckert DJ, White DP, Butler JP, Owens RL, Malhotra A, Wellman A: Acetazolamide improves loop gain but not the other physiological traits causing obstructive sleep apnoea. J Physiol 2012;590:1199–1211.
41 Wellman A, Malhotra A, Jordan AS, Stevenson KE, Gautam S, White DP: Effect of oxygen in obstructive sleep apnea: role of loop gain. Respir Physiol Neurobiol 2008;162:144–151.

David P. White, MD
4340 East Cedar Avenue
Denver, CO 80246 (USA)
E-Mail dpwhite@partners.org

Lin H-C (ed): Sleep-Related Breathing Disorders. Adv Otorhinolaryngol. Basel, Karger, 2017, vol 80, pp 17–21
DOI: 10.1159/000470683

The History of Sleep Surgery

Kathleen Yaremchuk • Laura Garcia-Rodriguez
Department of Otolaryngology – Head and Neck Surgery, Henry Ford Hospital, Detroit, MI, USA

Abstract

Snoring and the subsequent diagnosis of obstructive sleep apnea (OSA) was a life-threatening medical condition with no available treatment until the late 20th century. An early description of OSA was provided by Charles Dickens in his 1836 novel *Pickwick Papers* with the description of a "fat boy" who was thought to be lazy and always falling asleep but likely displayed hypersomnolence from OSA. It was not until 1976 that Ikematsu first described uvulopalatopharyngoplasty (UPPP) as an alternative surgical treatment of "snoring," with a reported cure rate of 81%. The only other surgical procedure for OSA was permanent tracheostomy, but patients suffered from social stigma from the visible stoma with skin flaps and complications such as tracheal granulomas and tracheitis. UPPP was introduced in the USA as an alternative to permanent tracheostomy by Fujita in 1981. Since then, multiple surgical approaches and combinations of approaches have surfaced, with variable success rates.

The diagnosis of sleep apnea stemmed from the constellation of symptoms found in the Pickwickian syndrome, a name derived from Charles Dickens' *Pickwick Papers* and his rotund character, Joe, who was often found to be sleeping (Fig. 1) [1]. Based on early reports of obstructive sleep apnea (OSA) in the 1950s, the disease was thought to be uncommon. Symptoms included hypersomnolence, rapid and shallow respirations, marked hypoventilation of pulmonary alveoli, and, in extreme cases, the requirement of positive-pressure ventilation [2]. Hundreds of cases were reported by the 1970s. In 1975, Carroll [3] described the differences between reports of the Pickwickian syndrome and OSA. Some patients had right-sided heart failure, respiratory acidosis, and no major abnormality in breathing, whereas others were hypersomnolent and had aberrations of respiration and sleep apneas secondary to upper airway obstruction. The latter patients had obstructions relieved by tracheostomy. In 1975, Sussman et al. [4] described 2 patients with Pickwickian syndrome with classical symptoms of periodic attacks of apnea, alveolar hypoventilation, hypersomnolence, hypertrophic tonsils, and "disturbing snoring at night."

Treatment of Snoring

Before the advent of sleep laboratories, most research focused on the treatment of snoring. In an address by Dr. Ian Robin in 1967, snoring was considered a nuisance or ridiculed, often afflicting the spouse but not the snorer [5]. Most patients did not present for evaluation since "so lit-

Fig. 1. Fat Joe from *Pickwick Papers*.

tle [was] done" [6]. In Robin's series of 200 patients, including children, he noted that the majority of the causes for snoring were adenoid hypertrophy (9%), tonsil and adenoid hypertrophy (19%), collapsed nasal ala (23%), deviated septum (25%), and hypertrophic turbinates. Snoring significantly improved with surgical intervention [6].

The sounds of snoring were speculated to be due to vibrations from the epiglottis, velum, cheeks, or nostrils, and were later confirmed with a nasopharyngoscope. Treatment was advocated but it was also considered that "isolation may be the only effective measure" for snoring [7]. Even in the 1970s, divorce was justified as a solution to snoring [8, 9]. Surgery was morbid and some chose alternative methods for snoring cessation, including sewing a marble to the upper part of a nightshirt as a reminder not to sleep supine [10] or using a cervical collar to maintain an extended neck position [9].

Continuous positive airway pressure (CPAP) was first used to treat pulmonary disease, such as pulmonary edema, but with endotracheal administration [11]. In 1975, Allen et al. [12] successfully introduced the application of a nontraumatic facemask for use with CPAP in newborns as a way to avoid endotracheal intubation. Treatment of OSA with CPAP was introduced by Sullivan et al. [13] in 1981, resulting in complete cessation of upper airway obstruction and a restful night's sleep. Since the introduction of CPAP, not everyone has gained access to the equipment and some find it cumbersome, so alternative treatment methods have been sought.

Surgical Procedures

In the 1950s, procedures for snoring included the resection of soft palate triangles just paramedian of the uvula, a concept invented by Schwartz [14]. This appeared to be the beginning of multiple procedures that used the same concept for reconstruction of the soft palate.

Amputation of the uvula was introduced in the 1960s, but only changed the pitch and not the occurrence of snoring [6]. Uvulectomy had been a tradition in Morocco for hundreds of years. Undertaken during infancy with a wooden spatula or a reed fork and a falciform bistoury, uvulectomy was meant to facilitate breastfeeding and speech. It was associated with the atrophy of regional lymphoid tissue and thus speculated that these infants were less prone to otitis media and adenotonsillitis [15]. Uvulectomy was not endemic to Morocco, but was widely practiced in African tribes. The procedure was performed by tribal healers based on the belief that it was preventative for throat and chest infections [16].

In the 1970s, tracheostomy became the standard treatment for Pickwickian syndrome, OSA, severe daytime somnolence with associated hypertension, and cardiac electrophysiologic changes. Most advocated performing a permanent stoma [17, 18]. However, in the 2 patients with Pickwickian syndrome studied by Sussman et al. [4], tonsillectomy ceased their symptoms with normalization of arterial blood gases 3 years postoperatively. Permanent tracheostomy had its complications of reported tracheal granulomas, tracheitis, and patients suffering social stigma from the permanent skin flap [19].

The Japanese practiced palatopharyngoplasties for many years before the procedure became popular in the USA [20]. Ikematsu [21] first proposed uvulopalatopharyngoplasty (UPPP) in 1964 in those afflicted with snoring (Fig. 2). His interest was first sparked in 1954 when a woman was divorced due to her loud snoring [22]. He advocated a morphological evaluation of the soft palate, uvula, pharynx, and the presence of snoring. Cephalometric analyses were performed to identify troublesome areas followed by a description of the surgical procedure and postoperative instructions. Treatment with surgery resulted in an 81% improvement of snoring [21].

64—434 日耳鼻

34. いびきの研究 第4報 いびきの一治療法

池 松 武 之 亮

1. はじめに

いびきは健康のシンボルでもなければ又正常なる睡眠呼吸音でもないという見地から，いびきの種類や性状によつてはその治療を必要とするものがあることを知つた．いびきの治療については久保村の軟口蓋パラフィン注入例，堀口の軟口蓋に高張食塩水注射例，林の後口蓋弓及口蓋垂切除法等の報告がある．私は今日まで280名のいびきを主訴とした者に接し，その中45名は鼻科的あるいは咽喉科的疾患の者であつたが，その他は何等いびきの原因と思われる疾患は見当らなつた．その中の152名に所謂いびきの手術的療法を行い，いささか知見を得たので報告する．

2. いびきの診断

いびきの治療にあたつては他の疾患と同様にその原因及び性状等を調べなければならない．

(1) 主訴に就いての検索

(2) 全身状態

(3) 上気道及び上気道周辺の疾患の有無

(4) 咽頭腔の形態及び軟口蓋や咽頭側索との相互関係

(5) 硬口蓋と軟口蓋との前後径の計測

(6) 後口蓋弓及び口蓋の計測

(7) いびきの聴取

以上のことを勘案して一時的抑制法か又は手術的療法にするかを決定する．いびきの抑制法に就いては他の機会に発表する予定である．

3. いびきの手術的療法の適応規準

いびきが鼻及咽頭等の疾患に発現されることは衆知の通りでこれらに就いて治療することは当然であるが，その疾患が治癒してもいびきが治らぬ時や疾患として認めるものがない時は次のような規準に従つて手術を決定している．

(1) いびきを聴取した結果それが口蓋垂の異常によると思われ，口蓋垂の長さが左右前口蓋弓の間隔の1/2.5倍以上長い時（テープコーダー供聴）

(2) 後口蓋弓が膜状を呈しその最大巾が5mm以上の時（テープコーダー）

(3) 正中線に於いて上咽頭腔の前後径が10mm以下の時（テープコーダー）

(4) 軟口蓋の正中線の前後径が硬口蓋のそれの1/1.5以上の時（テープコーダー）

(5) 咽頭側索炎のある時

(6) 充分なる開口にも関らず咽頭腔の直視不可能の時

4. 咽頭腔拡張術（いびきの手術法）スライド参照

(1) キシロカイン，アドレナリン咽頭塗布

(2) ノボカイン（0.5%）局所麻酔注射を右側の前後口蓋弓に行う．

(3) 前後口蓋弓の内端を口蓋垂の右端に沿つて約3mm小口蓋動脈の末端に達する出血点まで切り上げ，これを頂点として後口蓋弓に向け楔状に切除する．

(4) 前口蓋弓線の粘膜を2~3mmの幅で筋肉を傷つけることなく剝離除去し更に後口蓋弓粘膜で剝離除去する．

(5) 後口蓋弓から1又は2ヶ所に前口蓋弓縁直下を頂点として楔状に切除する．

(6) この弁状になつた後口蓋弓と前口蓋弓とを縫合する．

(7) 左側も同様に行う．

(8) 出血に対しては捻滅又は結紮する．

(9) 最後に正中線の口蓋垂根部に0.5ccのノボカイン注射を行い，前後2mmの間隔に小メスを刺入し，その部を小鉗子でとらえ粘膜と筋肉の小片を切除する．この結果創は卵円状となる．これを前後に縫合する．

(10) 口蓋垂は約0.8mm以上になればその尖端を前方より後上方に切除する．

(11) 咽頭側索炎のある場合はその粘膜を剝除する．

(12) 術後は24時間安静絶食

(13) 入院は3~7日

(14) 嚥下痛は5~7日間はあるが発熱は殆んど無し．

以上のようにして行えば軟口蓋の瘢痕による咽頭狭窄を防ぎ口蓋垂は前に傾き上咽頭腔は拡大する．

5. 沿革成績

(1) 鼻腔及び咽頭疾患のみの治療にていびきが治癒せしもの45名（16.1%）

(2) 咽頭腔拡張術を行つたもの152名男49名，女103名（54.3%）

更にこの沿革成績を示すと

野田市

(74)

Fig. 2. Ikematsu's proposal for uvulopalatopharyngoplasty in those afflicted with snoring [21].

Introduced in the USA by Fujita et al. [23] in 1981, UPPP was performed on 12 patients with OSA. Eight showed an improvement on polysomnogram and 9 had an improvement of subjective symptoms. The concept of the surgery was to enlarge the potential airspace in the oropharynx and cause stiffening of oropharyngeal tissues to avoid collapse and turbulent airflow. UPPP showed mixed results on polysomnograms, with some demonstrating up to 50% response rates with a variability in complications [24]. Patients would claim a significant subjective improvement in their symptoms despite objective changes on polysomnogram (Fig. 3).

In 1983, Cotton [25] presented the idea that UPPP should be used in conjunction with tracheostomy in patients with life-threating sleep apnea. That same year, Simmons et al. [26] presented the idea that the resection of superfluous palate could cure habitual snoring and treat at least 50% of OSA patients via palatopharyngoplasty. They warned of using the procedure in the morbidly obese or those with hypoventilation syndrome. The procedure was slightly different in that the uvula was simply excised, whereas it was repositioned in Fujita's approach. Simmons removed only as much soft palate as possible whilst avoiding velopharyngeal insufficiency (Fig. 4).

The earliest report of UPPP in a child was made in 1988, since most children are treated with adenotonsillectomy. The procedure was completed successfully in a 3-year-old with a large uvula and redundant soft palate, with a full resolution of symptoms at the 1-year follow-up [27].

Two phases of treatment have been suggested based on the complexity of OSA. Phase I, the conservative approach, includes UPPP, mandibular osteotomy with genioglossus advancement, hy-

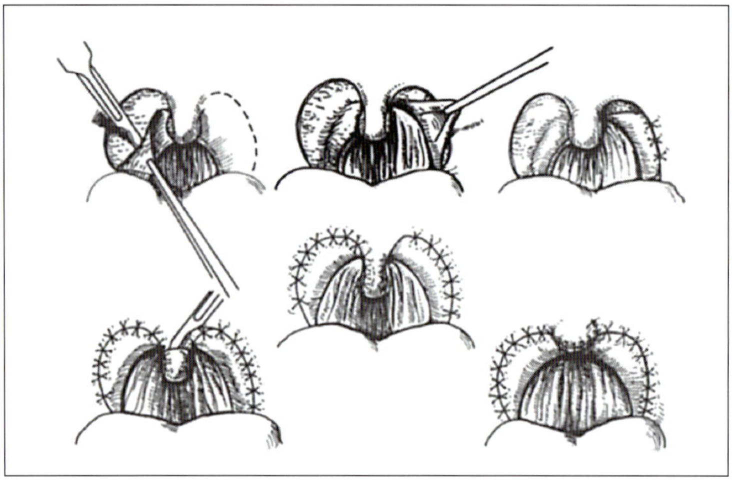

Fig. 3. Uvulopalatopharyngoplasty as described by Fujita et al. [23; with permission from SAGE Publications, Inc.] in 1981.

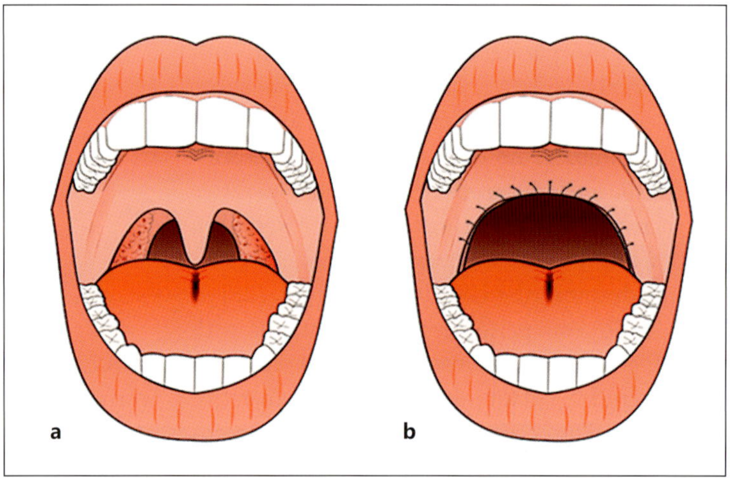

Fig. 4. Palatopharyngoplasty as described by Simmons et al. [25] in 1983. Preoperative (**a**) and original UPPP (**b**). The image is courtesy of Dr. Camacho (DrCamachoENT) and available at https://commons.wikimedia.org/wiki/File:UPPP._4_figures..png.

oid myotomy and suspension, tonsillectomy, and nasal surgery [28, 29]. Phase II procedures include maxillomandibular advancement osteotomy and temperature-controlled radiofrequency of the tongue base [29]. These phases are not clear-cut due to advancements in technology offering other means of treating OSA.

Based on the principle of altering the soft palate, the procedure introduced in 1990 by Kamami [30] using a CO_2 laser for vaporizations of the oropharynx is now commonly known as laser-assisted uvulopalatoplasty. Intended to "vaporize" vibrating tissues of the tonsillar pillars and soft palate, the procedure can be performed in 1 or more sessions as needed for the required results in an office setting. It was deemed "simple, bloodless, and painless," under the use of local anesthesia. However, the procedure was promoted for habitual snorers [30].

An extension of UPPP was presented by O'Leary and Millman [24] in 1991, in which they advocated resection of the palatopharyngeus and palatoglossus, increasing the area of the posterior oropharynx. Other procedures have been modeled after the principle of UPPP, such as snare uvulopalatoplasty, introduced by Weingarten [31] in 1991 for the treatment of habitual snoring. This procedure was meant to be performed in the clinic with local anesthesia using a snare.

Conclusion

Despite all of the advances made during these last few decades for OSA surgery, no single treatment is successful for all patients. As Guilleminault [32] said in 1998, "to offer only one of these [treatments] or to limit oneself to a single therapeutic approach is a long-term disservice."

References

1 Kryger MH: Fat, sleep, and Charles Dickens: literary and medical contributions to the understanding of sleep apnea. Clin Chest Med 1985;6:555–562.
2 Pickwickian syndrome. Br Med J 1958;2: 964.
3 Carroll D: Pickwickian syndrome, 20 years later. Trans Am Clin Climatol Assoc 1975;86:112–127.
4 Sussman D, Podoshin L, Alroy G: The Pickwickian syndrome with hypertrophy of tonsils: a re-appraisal. Laryngoscope 1975;85:565–569.
5 Snoring husbands. Chicago Tribune, July 16, 1968, p 14. http://archives.chicagotribune.com/1968/07/16/page/38/article/comic-5-no-title.
6 Robin I: Snoring [abridged]. Proc R Soc Lond 1968;61:575–582.
7 Snoring. Can Med Assoc J 1948;59:383–384.
8 Seiler ER: The incidence of snoring as a sleep problem in parents. J R Coll Gen Pract 1970;19:247.
9 Broomes EL: New thoughts on snoring prevention. J Natl Med Assoc 1982;74: 1139–1140.
10 Help a snorer. JAMA 1981;245:1729–1730.
11 Cheney FW Jr, Martin WE: Effects of continuous positive-pressure ventilation on gas exchange in acute pulmonary edema. J Appl Physiol 1971;30:378–381.
12 Allen LP, Blake AM, Durbin GM, Ingram D, Reynolds EO, Wimberley PD: Continuous positive airway pressure and mechanical ventilation by facemask in newborn infants. Br Med J 1975;4: 137–139.
13 Sullivan CE, Issa FG, Berthon-Jones M, Eves L: Reversal of obstructive sleep apnoea by continuous positive airway pressure applied through the nares. Lancet (London) 1981;1:862–865.
14 Heinberg CJ: A surgical procedure for the relief of snoring. Eye Ear Nose Throat Mon 1955;34:389.
15 Apffel CA: Uvulectomy, ethnic mutilation of prophylactic surgery: an oriental tale. JAMA 1965;193:164–165.
16 Sawe HR, Mfinanga JA, Ringo FH, Mwafongo V, Reynolds TA, Runyon MS: Morbidity and mortality following traditional uvulectomy among children presenting to the Muhimbili National Hospital Emergency Department in Dar es Salaam, Tanzania. Emerg Med Int 2015; 2015:108247.
17 Weitzman ED, Pollack CP, Borowiecki B: Hypersomnia-sleep apnea due to micrognathia: reversal by tracheoplasty. Arch Neurol 1978;35:392–395.
18 Fee WE Jr, Ward PH: Permanent tracheostomy: a new surgical technique. Ann Otol Rhinol Laryngol 1977;86:635–638.
19 Conway WA, Victor LD, Magilligan DJ Jr, Fujita S, Zorick FJ, Roth T: Adverse effects of tracheostomy for sleep apnea. JAMA 1981;246:347–350.
20 Simmons FB: Otolaryngology/head and neck surgery – important advances in clinical medicine: snoring. West J Med 1982;137:228.
21 Ikematsu T: Study of snoring: therapy [in Japanese]. J Jpn Otol Rhinol Laryngol Soc 1964;64:434–435.
22 Ikematsu A. YU, Niho M: The Ikematsu system for the diagnosis of snoring: a review of 39 years of clinical studies; in Togawa K, Katayama S, Hishikawa Y, Ohta Y, Horie T (eds): Sleep Apnea and Rhonchopathy. Basel, Karger, 1993.
23 Fujita S, Conway W, Zorick F, Roth T: Surgical correction of anatomic abnormalities in obstructive sleep apnea syndrome: uvulopalatopharyngoplasty. Otolaryngol Head Neck Surg 1981;89: 923–934.
24 O'Leary MJ, Millman RP: Technical modifications of uvulopalatopharyngoplasty: the role of the palatopharyngeus. Laryngoscope 1991;101:1332–1335.
25 Cotton RT: Uvulopalatopharyngoplasty. Arch Otolaryngol 1983;109:502.
26 Simmons FB, Guilleminault C, Silvestri R: Snoring, and some obstructive sleep apnea, can be cured by oropharyngeal surgery. Arch Otolaryngol 1983;109: 503–507.
27 Abdu MH, Feghali JG: Uvulopalatopharyngoplasty in a child with obstructive sleep apnea: a case report. J Laryngol Otol 1988;102:546–548.
28 Riley RW, Powell NB, Guilleminault C: Obstructive sleep apnea syndrome: a surgical protocol for dynamic upper airway reconstruction. J Oral Maxillofac Surg 1993;51:742–747; discussion 748–749.
29 Friedman M: Sleep Apnea and Snoring. Beijing, Elsevier, 2009.
30 Kamami YV: Laser CO_2 for snoring: preliminary results. Acta Otorhinolaryngol Belg 1990;44:451–456.
31 Weingarten C: Snare uvulopalatoplasty. Laryngoscope 1995;105:1033–1036.
32 Guilleminault C: Treatment of obstructive sleep apnea syndrome: a personal view based upon evaluation of over 4,000 patients. Schweiz Med Wochenschr 1988;118:1342–1347.

Kathleen Yaremchuk, MD
Department of Otolaryngology – Head and Neck Surgery
Henry Ford Hospital, 2799 West Grand Boulevard
Detroit, MI 48202 (USA)
E-Mail kyaremc1@hfhs.org

Lin H-C (ed): Sleep-Related Breathing Disorders. Adv Otorhinolaryngol. Basel, Karger, 2017, vol 80, pp 22–27
DOI: 10.1159/000470816

Advances in the Diagnosis of Obstructive Sleep Apnea: Drug-Induced Sleep Endoscopy

Madeline J.L. Ravesloot[a] · Linda Benoist[a, c] · Peter van Maanen[a] · Eric J. Kezirian[d] · Nico de Vries[a, b, e]

[a]Department of Otorhinolaryngology and Head and Neck Surgery, OLVG, and [b]Department of Oral Kinesiology, Academic Centre for Dentistry Amsterdam (ACTA), Amsterdam, and [c]Department of Otorhinolaryngology and Head and Neck Surgery, Erasmus University Medical Center, Rotterdam, The Netherlands; [d]USC Caruso Department of Otorhinolaryngology and Head and Neck Surgery, Keck School of Medicine of USC, Los Angeles, CA, USA; [e]Department of Otorhinolaryngology and Head and Neck Surgery, Faculty of Medicine and Health Sciences, Antwerp University Hospital (UZA), Edegem, Belgium

Abstract

Drug-induced sleep endoscopy was introduced in 1991 and has recently evolved into a pivotal instrument for patients in whom obstructive sleep apnea surgery is considered. Here, we discuss the indications, contraindications, technique, anesthesia, scoring systems, validity, and insights of the literature.

There are many – some fancy – synonyms in use for the same procedure, including sleep endoscopy, sleep nasendoscopy, somnoscopy, dormoscopy, drug-induced sedated sleep endoscopy, and drug-induced sleep endoscopy (DISE). The latter is the most often used today, in which the "S" stands for "sleep." The popularity of the procedure is on the rise. As many papers on DISE have been published over the last 3 years as were published in all previous years combined. This reflects accumulating evidence of the worth of DISE as a selection tool for surgery.

Although DISE is applied worldwide [1, 2], the reality is that its popularity is closely related to reimbursement. In countries where insurance companies do not reimburse DISE, ENT (ear, nose, and throat) specialists are forced to rely on clinical assessment only, for example by means of Friedman or Malampatti scoring. In countries where DISE is reimbursed, as in the Netherlands, it is regarded as wrong to schedule a patient for OSA surgery without performing DISE first.

The evaluation requires the pharmacologic induction of sedation and flexible fiber-optic endoscopy to visualize the upper airway obstruction and/or cause of snoring. DISE offers a dynamic evaluation of the upper airway during conditions that ideally mimic natural sleep, and also enables visualization of specific structures that contribute to upper airway obstruction.

Indications

DISE is employed as a diagnostic tool for patients with habitual snoring as well as those with obstructive sleep apnea (OSA) when surgery or mandibular advancement device (MAD) therapy is being considered as a treatment option. Furthermore, DISE can be applied to improve understanding of the anatomical basis for surgical, MAD, or CPAP (continuous positive airway pressure) failure, and to evaluate the full range of treatment alternatives [3].

DISE is not necessary if CPAP or weight loss are being considered, as visualization of the level of obstruction is not mandatory for these treatment modalities. It has to be realized that weight loss can improve certain unfavorable collapse patterns, for example a total concentric palatal collapse can reverse in a more favorable anterior-posterior (A-P) collapse.

A polysomnography (PSG) is pivotal in the workup of a patient as it is necessary to assess the severity, position dependency, and obstructive, mixed or central events. DISE on the other hand is crucial for the assessment of the site(s), levels, and configuration of obstruction. The interpretation of the polysomnography and DISE findings combined is paramount to surgical success. DISE helps to decide which surgical procedures can be applied, or might lead to the decision not to perform a certain type of surgery.

DISE can be performed in a standard fashion, while on indication certain passive maneuvers can be added. For instance, it is thought that by applying a jaw thrust/chin lift during DISE, one can predict the likelihood that a MAD would be effective [4]. On performing a chin-lift, one should bring the mandible forward by not more than 60–75% of the maximal mandibular protrusion. In our institute, if either surgery or MAD treatment are regarded as potential treatment options, DISE is no longer considered. In patients with only a MAD indication, it is debatable whether DISE, only to assess the effect of a chin lift, is worthwhile. Performing DISE costs money, but this is also true of producing a MAD, which can later prove to be ineffective.

In case positional therapy is considered as (part of) the approach, DISE can be performed in the lateral position [5, 6]. When a combination of mandibular repositioning appliance treatment with positional therapy is considered, a chin lift in the lateral position can be performed.

Contraindications

A high American Society of Anesthesiologists (ASA) score and propofol or midazolam allergies (albeit rare) are considered contraindications owing to the high risk. Due to the negative influence on treatment success, severe OSA and severe obesity are relative contraindications.

Technique

There are no standardized protocols concerning the DISE technique. The procedure can be carried out in the operating or endoscopy room and the sedation can either be administrated by an anesthesiologist or dedicated nurse anesthetist [7]. Pulse oximetry, heart rate, and blood pressure are closely monitored throughout the procedure, and it must be possible to administer oxygen if needed. Patients should remain nil per os before the procedure to reduce the risk of regurgitation and aspiration. The administration of atropine or other anticholinergic agents 30 min before starting the procedure can be considered to reduce salivation.

A topical anesthetic, with or without a decongestant, can be administered to 1 or both nostrils at least 20 min before starting the procedure, being careful not to over anesthetize the pharynx. This would increase the risk of aspiration and coughing, as well as incurring a potential interaction with the upper airway and breathing control [2].

The lights should be dimmed and the room quiet in order to minimize awaking stimuli. It is practical to be able to view the film of the flexible endoscopy on a screen and record it. With the help of a microphone, acoustic and visual signals can be recorded simultaneously.

Traditionally, the procedure is commenced with the patient in a supine position on an operating table or in a bed. The position should attempt to mimic sleeping habits at home (e.g., 1 or 2 pillows, with or without dentures). To gain added value, the body position should be easily changeable should one want to investigate the potential consequences of another position.

In 2 published reports, Safiruddin et al. [5, 6] evaluated the influence of different body positions on position-dependent OSA. It was found that the outcomes in the left and right lateral position are the same, and that turning the head only is almost equal to turning both the head and trunk. The clinical relevance is that instead of performing DISE in 5 different positions (head and trunk to the right, head only to the right, supine, head and trunk to the left and only the head to the left) and on indication performing a chin lift in all these 5 positions, turning the head only to 1 side suffices in patients with positional OSA.

Anesthesia

The drugs commonly used for DISE are propofol and/or midazolam. Some use propofol only while others use midazolam only. Others start with midazolam and continue with propofol.

Propofol has a rapid onset of action, is metabolized quickly, giving a fast recovery phase, and has a low incidence of postoperative nausea, vomiting, and headache. Propofol has the attraction of possessing a rapid onset of action and recovery period, with minimal side effects [8]. In addition, it allows for standardization and reproducibility between different operators.

A computerized target-controlled infusion system for propofol can be helpful, as well as a bispectral index score system for monitoring the depth of sedation; however, neither are compulsory. Bispectral index scores between 50 and 70 correlate most closely with the appropriate level of sedation for the DISE procedure.

The anesthetic depth is of key importance. The target depth of sedation is the transition from consciousness to unconsciousness (loss of response to verbal stimulation). Because individuals have different susceptibilities to propofol, the required dosage can vary widely. Detailed suggestions for drug dosages are reported in the European position paper on drug-induced sedation endoscopy (Table 1) [2].

Once the patient has reached a satisfactory level of sedation, a flexible endoscope (e.g., 3.5 mm), lubricated and applied with an anticondensation coating, is introduced into the nasal cavity. The nasal passage, nasopharynx, velum, tongue base, epiglottis, and larynx are observed. The levels of snoring and/or obstruction are assessed [9].

Scoring System

Various scoring systems, such as the VOTE classification and NOHL, are applied in clinical practice and used in the literature to report DISE findings. This hinders the scientific evaluation of DISE in individual centers and, just as importantly, the collection of data across multiple centers [2, 10–16]. The application of a universal scoring system would be of great significance, allowing the comparison of results across studies and increasing our knowledge and level of supporting evidence for whether or not DISE is indeed beneficial to the outcomes of existing and novel treatments for snoring and OSA. A scoring and classification system should include primary structures that contribute to upper airway obstruction, and the degree (severity) and configuration of upper airway obstruction (anteroposterior, circumferential or lateral) [2].

The philosophy behind the different scoring systems varies widely, and no consensus on a uni-

Table 1. Drug dosage table as presented in the European position paper on drug-induced sedation endoscopy by de Vito et al. [2]

	Drug dosage	
	midazolam	propofol
Midazolam	Bolus technique Starting dose: 0.03 mg/kg Observation 2–5 min If required, increase at a rate of 0.015–0.03 mg/kg	
Propofol		Target-controlled infusion Starting dose: 1.5–3.0 μg/kg If required, increase at a rate of 0.2–0.5 μg/kg
		Continuous pump Delivering dose: 50–100 ml/h
		Bolus technique Starting dose: 30–50 mg or 1 mg/kg If required, increase at a rate of 10–20 mg
Propofol and midazolam	Single bolus before administration of propofol: single starting dose 0.05 mg/kg or 1.5 mg	Target-controlled infusion Starting dose: 1.5–3.0 μg/kg If required, increase at a rate of 0.2–0.5 μg/kg

form system has been reached so far. Some score up to 7 levels of obstruction – including the nose, nasopharynx, and larynx – while others limit scoring to the collapsible segment only, as in VOTE. Some believe in very detailed semiquantitative scoring, while others have doubts on their reproducibility and are of the opinion that scoring in terms of no, partial and total collapse is already difficult enough. It is clear that with more levels to score, and with more detailed semiquantitative scoring, the inter- and intrarater variability will increase.

Validity

Previous studies reported that propofol did not change the respiratory pattern nor significantly influence the apnea-hypopnea index (AHI), but did interfere with the sleep architecture, specifically a reduction in rapid eye movement (REM) sleep in patients undergoing propofol-induced sedated endoscopy [17]. Respiratory and somnological parameters did not change significantly during diazepam-induced sedated endoscopy in comparison with natural sleep either, except for a small increase in the apnea index and a minor change in the duration of the longest apnea and REM sleep [9, 17].

The anesthetic depth is of key importance. The target depth of sedation is the transition from consciousness to unconsciousness (loss of response to verbal stimulation). Because individuals have differential susceptibilities to propofol, the required dosage can vary widely. A slow, stepwise induction is required to avoid oversedation. Deeper levels of sedation are associated with progressive decreases in upper airway dilator muscle tone and neuromuscular reflex activation, which both increase airway collapsibility, and the transition to unconscious sedation may be a closer approximation to natural sleep. Previous research using propofol has shown that the transition to unconsciousness is associated with changes in upper airway collapsibility (passive critical closing pressure), bispectral index score readings (based on frontal electroencephalography activity), and genioglossus muscle tone. Normal subjects have decreases in genioglossus tone to 10% of their maximum awake activity, which is half to a third of the level in non-OSA patients but greater than during REM sleep in controls and OSA patients [9]. While unconscious sedation under propofol may not be a perfect simulation of natural sleep, pharyngeal dilator muscle activity appears to lie somewhere between non-REM and REM sleep.

Insights from the Literature

The reliability of test-retest and the intra- and interobserver variability in DISE scoring are moderate to fair [16, 18–20]. Unsurprisingly, interobserver agreement is higher in ENT surgeons who are experienced with DISE [21].

There are no severe side effects, emergency situations, indications for endotracheal intubation, tracheotomy, or need of flumazenil administration described with DISE in the literature [9]. In case of oversedation, airway management with the use of positive pressure or, in rare cases, a laryngeal mask airway may be required.

Despite substantial heterogeneity in the literature, multilevel obstruction is a more common observation [18, 22–27]. The latter is associated with a higher AHI [23]. The presence of multilevel collapse, complete collapse, and base of tongue collapse is associated with a higher AHI [23, 24]. Palatal collapse is observed most frequently [22, 24, 27]. A complete concentric collapse (CCC), particularly at the palatal level, is associated with a higher BMI [28].

Just over 50% of 535 cases included in a recent systematic review had a different surgical treatment plan following DISE in comparison with awake examination [28]. The differences are most frequently associated with the hypopharyngeal and laryngeal structures. Of course, this gives no indication of whether DISE is associated with improved outcomes. Few studies consider the implications of DISE on surgical success and the studies that have been published are retrospective in nature [29, 30].

A CCC has repeatedly been associated with surgical failure in the literature. Patients with absence of CCC had a significant postoperative improvement in AHI following hypoglossal nerve stimulation therapy [31]. In a retrospective study, the presence of CCC was found to predict the occurrence of surgical failure [30]. CCC is associated with a high BMI, which in turn is associated with surgical failure.

As a rule, one can assume that unilevel collapse has better surgical outcomes than multilevel collapse, partial collapse is better than total collapse, and A-P collapse is more favorable than concentric collapse.

Conclusion

DISE is a valid addition to polysomnography and clinical assessment in an awake patient. Assuming the correct technique and performance, it is a reliable and safe tool to detect and analyze the phenomena of the upper airway during sleep, both visually and acoustically.

DISE allows for a targeted management of snoring or OSA in suitable individuals eligible for sleep surgery or MAD treatment by clarifying the underlying contributing anatomical sites. Further research is needed to evaluate the association between preoperative findings and surgical outcomes.

References

1 Croft CB, Pringle M: Sleep nasendoscopy: a technique of assessment in snoring and obstructive sleep apnoea. Clin Otolaryngol Allied Sci 1991;16:504–509.

2 de Vito A, Carrasco Llatas M, Vanni A, Bosi M, Braghiroli A, Campanini A, et al: European position paper on drug-induced sedation endoscopy (DISE). Sleep Breath 2014;18:453–465.

3 Kent DT, Rogers R, Soose RJ: Drug-induced sedation endoscopy in the evaluation of OSA patients with incomplete oral appliance therapy response. Otolaryngol Head Neck Surg 2015;153:302–307.
4 Johal A, Battagel JM, Kotecha BT: Sleep nasendoscopy: a diagnostic tool for predicting treatment success with mandibular advancement splints in obstructive sleep apnoea. Eur J Orthod 2005;27:607–614.
5 Safiruddin F, Koutsourelakis I, de Vries N: Upper airway collapse during drug induced sleep endoscopy: head rotation in supine position compared with lateral head and trunk position. Eur Arch Otorhinolaryngol 2015;272:485–488.
6 Safiruddin F, Koutsourelakis I, de Vries N: Analysis of the influence of head rotation during drug-induced sleep endoscopy in obstructive sleep apnea. Laryngoscope 2014;124:2195–2199.
7 Benoist L, de Vries N: Organization and logistics of drug-induced sleep endoscopy in a training hospital. Eur Arch Otorhinolaryngol 2015;272:2557–2559.
8 Glen JB: The development of 'Diprifusor': a TCI system for propofol. Anaesthesia 1998;53(suppl 1):13–21.
9 Hohenhorst W, Ravesloot MJL, Kezirian EJ, de Vries N: Drug-induced sleep endoscopy in adults with sleep-disordered breathing: technique and the VOTE classification system. Oper Tech Otolaryngol Head Neck Surg 2012;23:11–18.
10 Vicini C, de Vito A, Benazzo M, Frassineti S, Campanini A, Frasconi P, et al: The nose oropharynx hypopharynx and larynx (NOHL) classification: a new system of diagnostic standardized examination for OSAHS patients. Eur Arch Otorhinolaryngol 2012;269:1297–1300.
11 Pringle MB, Croft CB: A grading system for patients with obstructive sleep apnoea – based on sleep nasendoscopy. Clin Otolaryngol Allied Sci 1993;18:480–484.
12 Camilleri AE, Ramamurthy L, Jones PH: Sleep nasendoscopy: what benefit to the management of snorers? J Laryngol Otol 1995;109:1163–1165.
13 Kezirian E, Hohenhorst W, de Vries N: Drug-induced sleep endoscopy: the VOTE classification. Eur Arch Otorhinolaryngol 2011;268:1233–1236.
14 Bachar G, Nageris B, Feinmesser R, Hadar T, Yaniv E, Shpitzer T, et al: Novel grading system for quantifying upper-airway obstruction on sleep endoscopy. Lung 2012;190:313–318.
15 Koo SK, Choi JW, Myung NS, Lee HJ, Kim YJ, Kim YJ: Analysis of obstruction site in obstructive sleep apnea syndrome patients by drug induced sleep endoscopy. Am J Otolaryngol 2013;34:626–630.
16 Gillespie MB, Reddy RP, White DR, Discolo CM, Overdyk FJ, Nguyen SA: A trial of drug-induced sleep endoscopy in the surgical management of sleep-disordered breathing. Laryngoscope 2013;123:277–282.
17 Rabelo FAW, Braga A, Küpper DS, de Oliveira JAA, Lopes FM, de Lima Mattos PLV, et al: Propofol-induced sleep: polysomnographic evaluation of patients with obstructive sleep apnea and controls. Otolaryngol Head Neck Surg 2010;142:218–224.
18 Kezirian EJ, White DP, Malhotra A, Ma W, McCulloch CE, Goldberg AN: Interrater reliability of drug-induced sleep endoscopy. Arch Otolaryngol 2010;136:393–397.
19 Berry S, Roblin G, Williams A, Watkins A, Whittet HB: Validity of sleep nasendoscopy in the investigation of sleep related breathing disorders. Laryngoscope 2005;115:538–540.
20 Rodriguez-Bruno K, Goldberg AN, McCulloch CE, Kezirian EJ: Test-retest reliability of drug-induced sleep endoscopy. Otolaryngol Head Neck Surg 2009;140:646–651.
21 Vroegop AV, Vanderveken OM, Wouters K, Hamans E, Dieltjens M, Michels NR, et al: Observer variation in drug-induced sleep endoscopy: experienced versus nonexperienced ear, nose, and throat surgeons. Sleep 2013;36:947–953.
22 Salamanca F, Costantini F, Bianchi A, Amaina T, Colombo E, Zibordi F: Identification of obstructive sites and patterns in obstructive sleep apnoea syndrome by sleep endoscopy in 614 patients. Acta Otorhinolaryngol Ital 2013;33:261–266.
23 Vroegop AV, Vanderveken OM, Boudewyns AN, Scholman J, Saldien V, Wouters K, et al: Drug-induced sleep endoscopy in sleep-disordered breathing: report on 1,249 cases. Laryngoscope 2014;124:797–802.
24 Ravesloot MJ, de Vries N: One hundred consecutive patients undergoing drug-induced sleep endoscopy: results and evaluation. Laryngoscope 2011;121:2710–2716.
25 Abdullah VJ, Wing YK, van Hasselt CA: Video sleep nasendoscopy: the Hong Kong experience. Otolaryngol Clin North Am 2003;36:461–471, vi.
26 Bachar G, Feinmesser R, Shpitzer T, Yaniv E, Nageris B, Eidelman L: Laryngeal and hypopharyngeal obstruction in sleep disordered breathing patients, evaluated by sleep endoscopy. Eur Arch Otorhinolaryngol 2008;265:1397–1402.
27 Hamans E, Meeus O, Boudewyns A, Saldien V, Verbraecken J, van de Heyning P: Outcome of sleep endoscopy in obstructive sleep apnoea: the Antwerp experience. B-ENT 2010;6:97–103.
28 Certal VF, Pratas R, Guimaraes L, Lugo R, Tsou Y, Camacho M, et al: Awake examination versus DISE for surgical decision making in patients with OSA: a systematic review. Laryngoscope 2016;126:768–774.
29 Golbin D, Musgrave B, Succar E, Yaremchuk K: Clinical analysis of drug-induced sleep endoscopy for the OSA patient. Laryngoscope 2016;126:249–253.
30 Koutsourelakis I, Safiruddin F, Ravesloot M, Zakynthinos S, de Vries N: Surgery for obstructive sleep apnea: sleep endoscopy determinants of outcome. Laryngoscope 2012;122:2587–2591.
31 Vanderveken OM, Maurer JT, Hohenhorst W, Hamans E, Lin HS, Vroegop AV, et al: Evaluation of drug-induced sleep endoscopy as a patient selection tool for implanted upper airway stimulation for obstructive sleep apnea. J Clin Sleep Med 2013;9:433–438.

Madeline Ravesloot
Department of Otorhinolaryngology and Head and Neck Surgery
OLVG West, Jan Tooropstraat 164
NL–1061 AE Amsterdam (The Netherlands)
E-Mail m.j.l.ravesloot@olvg.nl

Lin H-C (ed): Sleep-Related Breathing Disorders. Adv Otorhinolaryngol. Basel, Karger, 2017, vol 80, pp 28–36
DOI: 10.1159/000470819

Novel Positional Devices for the Treatment of Positional Obstructive Sleep Apnea, and How This Relates to Sleep Surgery

Madeline J.L. Ravesloot[a] · Linda Benoist[a, c] · Peter van Maanen[a] · Nico de Vries[a, b, d]

[a]Department of Otorhinolaryngology and Head and Neck Surgery, OLVG, and [b]Department of Oral Kinesiology, Academic Centre for Dentistry Amsterdam (ACTA), Amsterdam, and [c]Department of Otorhinolaryngology and Head and Neck Surgery, Erasmus University Medical Center, Rotterdam, The Netherlands; [d]Department of Otorhinolaryngology and Head and Neck Surgery, Faculty of Medicine and Health Sciences, Antwerp University Hospital (UZA), Edegem, Belgium

Abstract

If untreated, obstructive sleep apnea (OSA) develops as a gradual progressive disease. In the early stage of the disease most patients with OSA are positional. The archetypical patient might progress from simple positional snoring via positional early-stage mild disease to less positional moderate and finally nonpositional severe OSA. At first, the apnea-hypopnea index (AHI) is high only in the supine position, and later is high in all sleeping positions. The phenomenon is reversible. After partial effective treatment, patients with severe OSA can reverse to less severe positional OSA or, in other words, the AHI drops more in the lateral position than in supine position. This has been shown for palatal surgery, multilevel surgery, bimaxillary osteotomies, and bariatric surgery. The absence or presence of positional dependency has a great influence on sleep surgery. First, the results of sleep surgery might be worse in positional patients. Second, the addition of positional therapy to sleep surgery might improve the overall outcome and, as such, enhance the indication of sleep surgery as an alternative to continuous positive airway pressure and mandibular advancement device treatment.

In approximately 56–75% of patients with obstructive sleep apnea (OSA), the frequency of apneas and hypopneas is influenced by body position [1–6]. Usually, but not necessarily, the worst sleeping position is supine. The prevalence of position-dependent OSA (POSA) is higher in Asians [5, 6]. POSA is defined as having twice as many respiratory events in the supine sleeping position compared to the nonsupine position [2]. The majority of patients with POSA (70–80%) have mild

OSA. POSA is less common in moderate disease than in mild disease, and rare in severe OSA. Before OSA has become manifest, in snoring patients with an apnea-hypopnea index (AHI) below 5, position dependency occurs in 68% [1, 7]. There is an inverse relationship between POSA and body mass index. In a recent paper, 60–70% of patients on the waiting list for bariatric surgery had OSA, while 34% had POSA [8].

Patients with POSA also tend to have a smaller neck circumference [9]. Various studies have suggested that positional patients in comparison to nonpositional patients have a more backward positioning of the lower jaw, lower facial height, longer posterior airway space measurements, and a smaller volume of lateral pharyngeal wall tissue and, therefore, greater lateral diameter of the upper airway [10–14].

In theory, a great proportion of patients with OSA, in particular patients with mild or moderate disease severity and a low body mass index, could benefit from positional therapy (PT). PT aims at preventing patients from sleeping in the worst sleeping position [1]. Traditionally, various techniques have been tested, such as positional alarms. These positional alarms had the significant limitation that they woke patients up, and for this reason are no longer used.

The majority of earlier studies on PT apply the so-called tennis ball technique (TBT), which, including variations on the theme, involves a bulky mass strapped to the patient's back. Even though the TBT is simple and cheap, as well as effective in reducing the AHI, results remain unsatisfactory. Backache, discomfort, and no improvement or even a deterioration of sleep quality or daytime alertness have been responsible for poor compliance and subsequent disappointing long-term results [15]. Compliance rates reported in the literature range from 40% in the short term to 10% in the long term. Arousals whilst turning from 1 lateral position to the other, thereby disturbing sleep architecture and quality, as well as the inherent discomfort account for these results [15–17]. TBT techniques have been analyzed in all variations, including tennis balls, squash balls, inflatable masses, and shark fins, but the major drawback of low compliance holds true for all variations on the TBT theme. Consequently, the TBT should no longer be used.

Recent developments have seen the introduction of a new generation of PTs. These include small smart devices, attached to either the chest or neck, which prevent the patient from adopting the supine position through a subtle vibrating stimulus [15]. Encouraging data suggest that this simple therapy successfully prevents patients with POSA from adopting the supine position without negatively influencing sleep efficiency, as well as achieving good adherence. This new generation of PTs, which are patient friendly, reversible, without important side effects, both simple and relatively inexpensive, shows promise as a standalone treatment or as an additional measure to increase the success rate of other therapies. PT can be combined with surgery, mandibular advancement device (MAD) therapy, (surgical) weight loss and, to a lesser extent, continuous positive airway pressure (CPAP). Here we will focus on PT alone, PT combined with MAD, and PT combined with surgery.

Chest-Worn and Neck-Worn Devices

The position monitoring and supine alarm device (Fig. 1) and the Sleep Position Trainer (SPT; Fig. 2) are both small, lightweight, battery-powered apparatus. They weigh 50 and 25 g, and have dimensions of 80 × 40 × 20 and 72 × 35 × 10 mm, respectively. Both are fastened to the chest with a strap so that the device is located on the sternum. The SPT is placed in a pocket of the strap, which is closed with a Velcro tab.

The SPT described by Bignold et al. [18] gives off a vibration if the supine position is detected

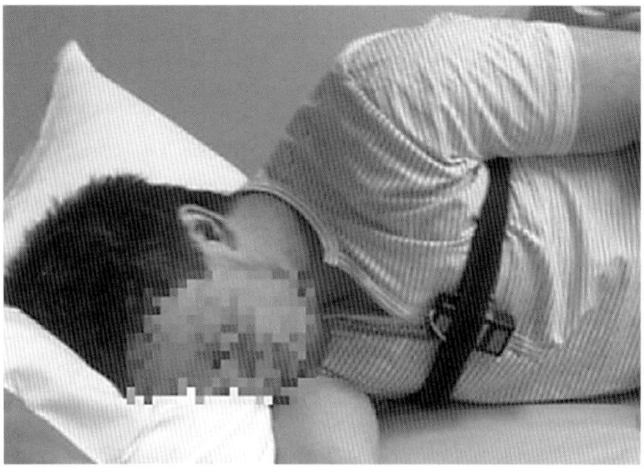

Fig. 1. The position monitoring and supine alarm device.

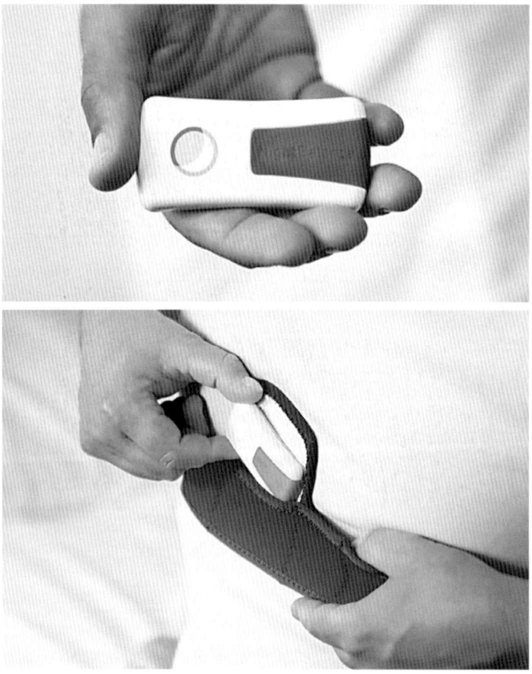

Fig. 2. The Sleep Position Trainer.

for 5 consecutive seconds with no subsequent turning movement. The stimulus increases gradually until a nonsupine position is detected. If no reaction is monitored, the vibrations are paused and reinitiated after 2 min. When the reaction time increases, the vibrating pattern changes. Both devices have the capacity to store and register data such as sleeping posture, which can be uploaded through custom software via a USB connection and personal computer [18–22]. The vibrating stimulus generated by both devices if the supine position is identified is measured using internal position-sensitive tilt switches [18] or a 3-dimensional digital accelerometer [19–22].

The original primitive neck-worn device (Fig. 3) described by van Maanen et al. [23] consisted of a small vibrating apparatus (30 × 30 × 10 mm) powered by 3 small batteries. The position sensor provided a vibrating stimulus with a delay of 10 s after the supine position was detected, causing the device to vibrate with gradually incremental strength until a different sleeping position was detected, in which case the vibrations ceased immediately. The small device was worn secured to the skin of the neck with hypoallergenic adhesive tape and connected to the polysomnography system. The use of this system has since been abandoned.

The Neck Position Therapy Device as described by Levendowski et al. [24] is a small battery-powered device weighing 44 g (55 × 38 × 16 mm) affixed on the back of the neck with an adjustable non-latex silicone rubber strap secured by a magnetic clasp (Fig. 4). Using a 3-dimensional digital accelerometer to detect the sleeping position, a vibrating stimulus is provided by two 1-G haptic motors if the supine position is adopted. Positional feedback is modulated by setting the number of motors to be excited (1 or both) and varying the duration of the motor(s) excitation. Feedback is initiated at a very low frequency/duration, and gradually increased until the user moves from the supine position. At any given intensity level, the feedback is repeated 6 times with an inter-feedback interval of 2 s. This device is also capable of storing data.

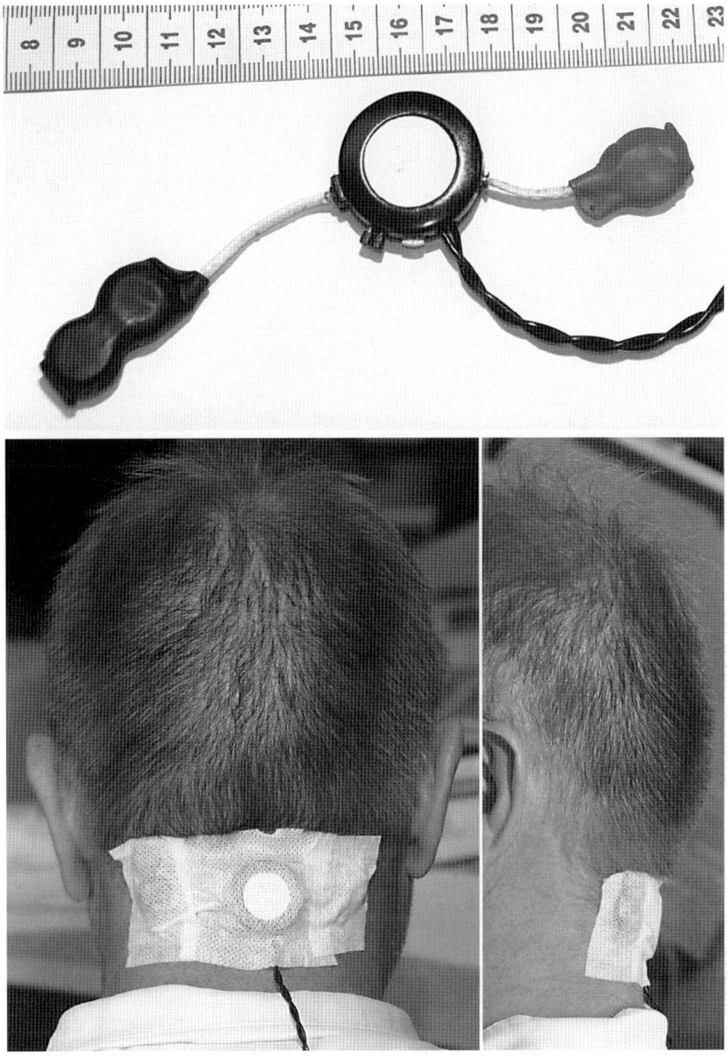

Fig. 3. A neck-worn positional therapy device. The middle part (ring structure) of the apparatus shown consists of a small vibrating motor (such as that used in cell phones) and a position sensor. Three small round batteries (2 positioned on the left, 1 on the right) are connected via the white cables. The braided black cables connect the device to the polysomnography system.

Randomized Controlled Trials

A summary of randomized controlled trials and the results on PT as the only treatment or combined with MAD treatment can be found in Tables 1 and 2, respectively. Two articles have reported randomized crossover trials comparing active versus inactive PT. Bignold et al. [18] evaluated the efficacy of a chest-worn device (Fig. 1) in 15 patients fulfilling the following criteria: overall AHI ≥15, supine AHI at least double the nonsupine AHI, ≥20 min of sleep in supine and nonsupine postures, and a nonsupine AHI <20. The mean baseline AHI (24.1) was reduced by approximately 45% with active treatment.

van Maanen et al. [23] performed a study evaluating the efficacy of a neck-worn device (Fig. 3) that included 30 patients with POSA. Assigned at random, in 1 of the 2 test recordings the device was active, while in the other it was inactive, and

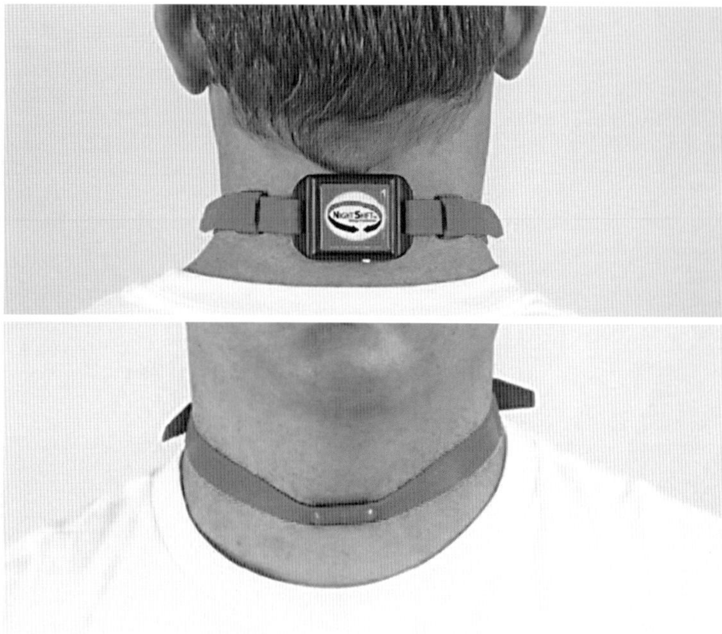

Fig. 4. Neck position training device.

patients were blinded to the chosen activity state of the device. In the treatment group there was a statistically significant mean drop in AHI from 27.7 to 12.8 and 23% developed an overall AHI <5. The device did not have a significant effect on sleep efficiency.

Without taking the compliance into consideration, Eijsvogel et al. [19] observed that the SPT (Fig. 2) and TBT were equally effective in reducing respiratory indices in 55 patients in a randomized controlled trial. Both therapies reduced the supine sleep position to a median of 0%. An AHI <5 was achieved in 43 and 68% of the patients treated with TBT and SPT, respectively (p = 0.087). Compliance, defined as ≥4 h/night and ≥5 nights/week, was 75.9% for the SPT and 42.3% for TBT users (p = 0.01). There were more dropouts in the TBT group (5 vs. 2). The reasons for dropout were shoulder/back pain, inability to adopt a supine position, or vibrating noise. A mean disease alleviation of 48.6 and 70.5% for TBT and SPT, respectively, was observed (p = 0.005). Objective sleep quality parameters and awakenings improved more in the SPT group, as did the Quebec Sleep Questionnaire scores.

Mandibular Advancement Device and Positional Therapy Combinations

Dieltjens et al. [22] evaluated the additional effect of SPT in patients with residual POSA during MAD therapy in an randomized controlled trial performed in 20 patients. One arm received SPT prior to SPT with MAD, whilst the second group received SPT with MAD, followed by SPT alone. The SPT reduced the time spent in a supine sleeping position compared to baseline and MAD therapy. Both MAD and SPT were individually effective in reducing the overall AHI significantly when compared to baseline from a median 20.9 at baseline to 11.0 and 12.8 with MAD or SPT, respectively. When treated with a combination of SPT with MAD, the AHI was further reduced to

Table 1. Summary of included randomized controlled trial studies

First author	Year	Design	Intervention	Location of device	Control	Sample size, *n*	Loss to follow-up, *n*	BMI	Age, years	Male, %	Mean baseline AHI	Questionnaires and scales	Follow-up, weeks
Bignold [18]	2011	Randomized crossover trial	Position monitoring and supine alarm device	Chest	Inactive treatment	16	1	28.8±2.5	58.2±13.9	86.7	24.1±10.5	Snoring scale score questionnaire	3
van Maanen [23]	2012	Randomized controlled single-blind, crossover trial	Neck-worn vibrating apparatus	Neck	Inactive treatment	30	0	27.7	48.0±9.5	60.0	27.7±3.6	–	–
Dieltjens [22]	2015	Randomized controlled trial	SPT	Chest	MAD; SPT + MAD	20	0	26.4±3.0	52.5±10.5	57.9	20.9 (17.0; 34.0)[a]	–	–
Eijsvogel [19]	2015	Randomized trial	SPT	Chest	TBT	55	I: 2 C:5	I: 27.6±4.5 C: 26.8±3.0	I: 50.1±10.6 C: 50.7±12.2	I: 79.3 C: 84.6	I: 11.4±4.9 C: 13.1±9.1	ESS, QSQ	4

AHI, apnea-hypopnea index; BMI, body mass index; C, control; ESS, Epworth sleepiness scale; I, intervention; MAD, mandibular advancement device; QSQ, Quebec sleep questionnaire; SPT, sleep position trainer; TBT, tennis ball technique. [a] Median (quartile 1; quartile 3).

Table 2. Overview of the results of the included randomized controlled trial studies

	Bignold [18]		van Maanen [23]		Dieltjens [22]		Eijsvogel [19]	
	intervention active PT	control inactive PT	intervention active PT	control inactive PT	intervention active PT	control SPT + MAD	intervention active PT	control TBT
Baseline AHI	24.1±10.5	24.1±10.5	27.7±2.4	27.7±2.4	20.9 [17.0; 34.0]	20.9 [17.0; 34.0]	11.4±4.9	13.1±9.1
AHI with treatment	13.7±1.1	25.0±1.7	12.8±2.2	23.5±2.6	12.8 [3.9; 17.9]	5.5 [3.4; 7.2]	3.9 (0.4–30.8)	5.8 (0.2–23.1)
TST supine position baseline, %	36.4±20.6	36.4±20.6	40±3.5	40±3.5	31.9 [15.4; 52.4]	31.9 [15.4; 52.4]	27.9 (10.3–77.6)	31.1 (10.7–73.7)
TST supine position with treatment, %	0.4±0.3	19.3±4.3	19±4.1	4.0±4.5	0.0 [0.0; 1.1]	0.0 [0.0; 1.0]	0.0 (0.0–67.6)	0.0 (0.0–38.9)
Baseline AHI in supine position	51.3±23.3	51.3±23.3	59.7±3.6	59.7±3.6	39.1 [26.4; 58.2]	39.1 [26.4; 58.2]	30.7±15.3	37.3±24.0
AHI in supine position with treatment	–	–	12.5±3.1	45.0±4.8	0.0 [0.0; 11.5]	0.0 (0.0; 22.7]	0.0 (0.0–64.2)	0.0 (0.0–116)
Baseline AHI in nonsupine position	9.7±3.9	9.7±3.9	6.7±1.2	6.7±1.2	11.1 [6.3; 26.1]	11.1 [6.3; 26.1]	3.9 (0.5–13.0)	3.3 (0.0–13.7)
AHI in nonsupine position with treatment	–	–	11.2±2.2	13.4±2.7	12.8 [3.8; 17.2]	4.8 [2.0; 6.0]	3.6 (0.4–30.9)	5.0 (0.3–14.2)
Baseline sleep efficiency, %	81.4±9.2	81.4±9.2	91.9±1.4	91.9±1.4	85.1 [79.4; 87.2]	85.1 [79.4; 87.2]	83.5 (63.0–92.2)	87.6 (58.0–93.7)
Sleep efficiency with treatment, %	–	–	88.3±1.8	89.9±1.6	84.6 [80.6; 87.9]	84.2 [79.0; 89.7]	82.8±9.8	81.3± 8.7
Baseline arousal index	17.3±6.9	17.3±6.9	9.0±1.1	9.0±1.1	10.1 [6.6; 15.0]	10.1 [6.6; 15.0]	10.8±5.8	12.5±6.2
Arousal index with treatment	–	–	6.8±0.7	9.9±1.0	19.5 [11.2; 24.3]	11.7 [6.7; 13.6]	8.1 (3.8–33.3)	9.6 (0.5–19.9)
Baseline ODI	–	–	11.5±1.7	11.5±1.7	7.7 [6.6; 16.5]	7.7 [6.6; 16.5]	9.9±5.0	10.9±7.7
ODI with treatment	3.4±0.2	5.5±0.3	4.6±1.1	9.7±1.8	2.6 [1.0; 4.6]	18.8 [1.0; 3.0]	4.4 (0.5–33.8)	5.4 (0.4–15.1)

Data are presented as the median (range) or median [interquartile range]. AHI, apnea-hypopnea index; BMI, body mass index; MAD, mandibular advancement device; ODI, oxygen desaturation index; PT, positional therapy; RCT, randomized controlled trial; SPT, sleep position trainer; TBT, tennis ball technique; TST, total sleep time.

5.5. This was statistically significantly lower when compared with baseline, MAD alone, and SPT alone. The authors concluded that the combination of SPT and MAD leads to a higher therapeutic efficacy in patients with residual POSA during MAD therapy in comparison to either therapy alone.

Positional Therapy and Head Position

A recent study demonstrated the importance of head position, separate from trunk position, as an important parameter to measure in patients with POSA [25]. The authors concluded that in a quarter of all OSA patients the head position is an important additional factor next to the position of the trunk. In these patients, the AHI calculated whilst the head is in a supine position was higher than the AHI determined while the trunk was in a supine position. Furthermore, 6.5% of the patients were not position dependent based on the position sensor on the trunk and were classified as only head-supine dependent. This finding may have consequences for the manner in which we diagnose and treat OSA patients in the future.

Sleep Position and Palatal Surgery

Four studies have addressed the issue of the influence of sleep position after palatal surgery for OSA. Katsantonis et al. [26] evaluated the effect of uvulopalatopharyngoplasty (UPPP) as related to sleep posture. The authors found a significant improvement of AHI in the lateral sleep position following UPPP and suggested that additional PT could significantly improve the response to treatment with UPPP.

Lee et al. [27] evaluated the effect of sleep position on surgical outcome. Patients who underwent UPPP were categorized into 4 groups. They showed that the failure group had a higher proportion of POSA compared to the other groups. A second paper from the same investigators indicated that UPPP is a successful treatment for obstructive events occurring in the lateral sleep position, especially in patients without positional dependency [28].

van Maanen et al. [29] analyzed the results of 130 patients who underwent either isolated tongue base or multilevel surgery for OSA. They concluded that successful and nonsuccessful surgical outcomes could not be explained by variations in the percentages of supine sleep position and that sleep position is not a confounding factor on surgical outcomes in tongue base surgery. However, a later study by van Maanen et al. [30] showed that, in patients who previously underwent UPPP or zetapalatoplasty with or without radiofrequency ablation of the tongue base, the reduction in AHI was significantly higher in nonpositional OSA patients compared to POSA patients. They suggested applying PT after surgery or even to start with PT as a monotherapy prior to surgery.

All these studies confirm that the effect of UPPP is most successful in decreasing AHI in the lateral sleep position. Additional postoperative PT could potentially improve the treatment outcome.

Positional Therapy after Maxillomandibular Advancement

In patients with severe or extreme OSA and CPAP failure, bimaxillary osteotomies or maxillomandibular advancement (MMA) can be considered. Because of its considerable morbidity, MMA is mostly reserved for severe to extreme cases, often after less aggressive surgeries have failed. The success rates of MMA are very high, reported to be between 80 and 90%. A new finding is that MMA failure can be due to the fact that after surgery the AHI in the supine sleeping position remains much higher as compared to the nonsupine position; here again, severe nonpositional OSA has reversed into less severe POSA. Postoperative residual POSA can in theory be a good indication for additional PT.

In selected patients with severe to extreme OSA, MMA can be indicated both as a primary surgical modality and in the case of a failure of multilevel surgery. We observed that in a number of patients in whom MMA was performed, the total AHI decreased, but not sufficiently to fulfil the criteria for success [31]. Since MMA is often the ultimum refugium, and taking into account its considerable morbidity, this is a very disappointing outcome. However, on closer inspection, in many of these patients only the AHI in a supine sleeping position remained high, whereas the AHI in other sleeping positions improved significantly. These patients had reversed from severe nonpositional OSA to less severe POSA. In a group of 42 patients who underwent MMA surgery, 9 were diagnosed with POSA after surgery. The diagnosis in these cases was mainly due to an AHI that was still high (in 4 cases still above 30) in the supine sleeping position, i.e., residual POSA. The mean supine AHI decreased from 74.5 to 36.9, representing a reduction of 50.5%. The mean AHI in lateral sleeping positions decreased even more; from 42.4 to 7.0 (83.5%) in the right sleeping position, and from 52.7 to 4.8 (90.9%) in the left sleeping position. The results of this study indicate that persistent OSA after MMA surgery can be caused by residual OSA in the supine position.

Positional Therapy and Bariatric Surgery

Morong et al. [32] found that patients with severe OSA who underwent bariatric surgery reversed into POSA postoperatively with a lower disease severity.

Conclusion

All studies published on new-generation PT have shown that it is effective in reducing the AHI. Furthermore, it is simple, relatively inexpensive, well tolerated, and reversible. Not only is PT efficacious, it is also effective, due to its high compliance. Further high-quality studies are needed to confirm the promising role of PT as a single or as a combination treatment modality. Regarding surgery, the same pattern might occur after UPPP, multilevel surgery, weight loss because of bariatric surgery, and MMA. The AHI does decrease in a lateral sleeping position, but sometimes not sufficiently in a supine sleeping position. Postoperative residual POSA can be a good indication for additional PT.

References

1 Ravesloot MJ, Frank MH, van Maanen JP, Verhagen EA, de Lange J, de Vries N: Positional OSA part 2: retrospective cohort analysis with a new classification system (APOC). Sleep Breath 2016;20: 881–888.

2 Cartwright RD: Effect of sleep position on sleep apnea severity. Sleep 1984;7: 110–114.

3 Oksenberg A, Silverberg DS, Arons E, Radwan H: Positional vs nonpositional obstructive sleep apnea patients: anthropomorphic, nocturnal polysomnographic, and multiple sleep latency test data. Chest 1997;112:629–639.

4 Richard W, Kox D, den Herder C, Laman M, van Tinteren H, de Vries N: The role of sleep position in obstructive sleep apnea syndrome. Eur Arch Otorhinolaryngol 2006;263: 946–950.

5 Mo JH, Lee CH, Rhee CS, Yoon IY, Kim JW: Positional dependency in Asian patients with obstructive sleep apnea and its implication for hypertension. Arch Otolaryngol Head Neck Surg 2011;137: 786–790.

6 Teerapraipruk B, Chirakalwasan N, Simon R, Hirunwiwatkul P, Jaimchariyatam N, Desudchit T, et al: Clinical and polysomnographic data of positional sleep apnea and its predictors. Sleep Breath 2012;16:1167–1172.

7 Benoist LB, Morong S, van Maanen JP, Hilgevoord AA, de Vries N: Evaluation of position dependency in non-apneic snorers. Eur Arch Otorhinolaryngol 2014;271:189–194.
8 Morong S, Benoist LB, Ravesloot MJ, Laman DM, de Vries N: The effect of weight loss on OSA severity and position dependence in the bariatric population. Sleep Breath 2014;18:851–856.
9 Mador MJ, Kufel TJ, Magalang UJ, Rajesh SK, Watwe V, Grant BJ: Prevalence of positional sleep apnea in patients undergoing polysomnography. Chest 2005;128:2130–2137.
10 Saigusa H, Suzuki M, Higurashi N, Kodera K: Three-dimensional morphological analyses of positional dependence in patients with obstructive sleep apnea syndrome. Anesthesiology 2009; 110:885–890.
11 Chang ET, Shiao GM: Craniofacial abnormalities in Chinese patients with obstructive and positional sleep apnea. Sleep Med 2008;9:403–410.
12 Walsh JH, Leigh MS, Paduch A, Maddison KJ, Armstrong JJ, Sampson DD, et al: Effect of body posture on pharyngeal shape and size in adults with and without obstructive sleep apnea. Sleep 2008; 31:1543–1549.
13 Soga T, Nakata S, Yasuma F, Noda A, Sugiura T, Yatsuya H, et al: Upper airway morphology in patients with obstructive sleep apnea syndrome: effects of lateral positioning. Auris Nasus Larynx 2009;36:305–309.
14 Pevernagie DA, Stanson AW, Sheedy PF 2nd, Daniels BK, Shepard JW Jr: Effects of body position on the upper airway of patients with obstructive sleep apnea. Am J Respir Crit Care Med 1995;152: 179–185.
15 Ravesloot MJ, van Maanen JP, Dun L, de Vries N: The undervalued potential of positional therapy in position-dependent snoring and obstructive sleep apnea – a review of the literature. Sleep Breath 2013;17:39–49.
16 Bignold JJ, Deans-Costi G, Goldsworthy MR, Robertson CA, McEvoy D, Catcheside PG, et al: Poor long-term patient compliance with the tennis ball technique for treating positional obstructive sleep apnea. J Clin Sleep Med 2009;5: 428–430.
17 Oksenberg A, Silverberg D, Offenbach D, Arons E: Positional therapy for obstructive sleep apnea patients: a 6-month follow-up study. Laryngoscope 2006;116:1995–2000.
18 Bignold JJ, Mercer JD, Antic NA, McEvoy RD, Catcheside PG: Accurate position monitoring and improved supine-dependent obstructive sleep apnea with a new position recording and supine avoidance device. J Clin Sleep Med 2011; 7:376–383.
19 Eijsvogel MM, Ubbink R, Dekker J, Oppersma E, de Jongh FH, van der Palen J, et al: Sleep position trainer versus tennis ball technique in positional obstructive sleep apnea syndrome. J Clin Sleep Med 2015;11:139–147.
20 van Maanen JP, Meester KA, Dun LN, Koutsourelakis I, Witte BI, Laman DM, et al: The sleep position trainer: a new treatment for positional obstructive sleep apnoea. Sleep Breath 2013;17:771–779.
21 van Maanen JP, de Vries N: Long-term effectiveness and compliance of positional therapy with the sleep position trainer in the treatment of positional obstructive sleep apnea syndrome. Sleep 2014;37:1209–1215.
22 Dieltjens M, Vroegop AV, Verbruggen AE, Wouters K, Willemen M, De Backer WA, et al: A promising concept of combination therapy for positional obstructive sleep apnea. Sleep Breath 2015;19: 637–644.
23 van Maanen JP, Richard W, van Kesteren ER, Ravesloot MJ, Laman DM, Hilgevoord AA, et al: Evaluation of a new simple treatment for positional sleep apnoea patients. J Sleep Res 2012;21: 322–329.
24 Levendowski DJ, Seagraves S, Popovic D, Westbrook PR: Assessment of a neck-based treatment and monitoring device for positional obstructive sleep apnea. J Clin Sleep Med 2014;10:863–871.
25 van Kesteren ER, van Maanen JP, Hilgevoord AA, Laman DM, de Vries N: Quantitative effects of trunk and head position on the apnea hypopnea index in obstructive sleep apnea. Sleep 2011; 34:1075–1081.
26 Katsantonis GP, Miyazaki S, Walsh JK: Effects of uvulopalatopharyngoplasty on sleep architecture and patterns of obstructed breathing. Laryngoscope 1990; 100:1068–1072.
27 Lee CH, Shin HW, Han DH, et al: The implication of sleep position in the evaluation of surgical outcomes in obstructive sleep apnea. Otolaryngol Head Neck Surg 2009;140:531–535.
28 Lee CH, Kim SW, Han K, et al: Effect of uvulopalatopharyngoplasty on positional dependency in obstructive sleep apnea. Arch Otolaryngol Head Neck Surg 2011;137;675–679.
29 van Maanen JP, Ravesloot MJL, de Vries N: Positional therapy and tongue base surgery; in de Vries N, Ravesloot MJL, van Maanen JP (eds): Positional Therapy in Obstructive Sleep Apnea. Cham, Springer, 2015, pp 311–320.
30 van Maanen JP, Witte BI, de Vries N: Positional therapy and palatal surgery; in de Vries N, Ravesloot MJL, van Maanen JP (eds): Positional Therapy in Obstructive Sleep Apnea. Cham, Springer, 2015, pp 297–310.
31 Benoist L, de Ruier MHT, de Lange J, de Vries N: Residual POSA after maxillo-mandibular advancement in patients with severe OSA; in de Vries N, Ravesloot MJL, van Maanen JP (eds): Positional Therapy in Obstructive Sleep Apnea. Cham, Springer, 2015, pp 321–330.
32 Morong S, Benoist L, de Vries N: Positional OSA in the morbidly obese and the effect of weight loss on OSA severity; in de Vries N, Ravesloot MJL, van Maanen JP (eds): Positional Therapy in Obstructive Sleep Apnea. Cham, Springer, 2015, pp 185–192.

Madeline Ravesloot
Department of Otorhinolaryngology and Head and Neck Surgery
OLVG West, Jan Tooropstraat 164
NL–1061 AE Amsterdam (The Netherlands)
E-Mail m.j.l.ravesloot@olvg.nl

Lin H-C (ed): Sleep-Related Breathing Disorders. Adv Otorhinolaryngol. Basel, Karger, 2017, vol 80, pp 37–40
DOI: 10.1159/000470821

Updated Concepts on Treatment Outcomes for Obstructive Sleep Apnea

Victor Certal[a,b] • Robson Capasso[c]

[a]Department of Otorhinolaryngology and Sleep Medicine Center, Hospital CUF, Porto and Centro Hospitalar de Entre o Douro e Vouga EPE (CHEDV), and [b]Centre for Research in Health Technologies and Information Systems (CINTES), University of Porto, Porto, Portugal; [c]Department of Otolaryngology – Head and Neck Surgery, Division of Sleep Surgery, Stanford Hospital and Clinics, Stanford, CA, USA

Abstract

Obstructive sleep apnea is a highly prevalent and poorly understood condition that not only leads to significant individual patient health damage, but carries a large societal cost secondary to excessive healthcare utilization, motor-vehicle accidents, and work absenteeism. While a significant fund of knowledge about its damaging effects on the cardiovascular system and neural pathways has been gathered over the past few decades, the translation to direct patient care still presents numerous challenges. With the exception of clear issues such as socially unacceptable snoring and breathing irregularities, the exact cause and effect relationship between airway narrowing, hypoxemia, and sleep fragmentation with subjective sleep-related complaints and neurocognitive symptoms is less clear. Besides better-defined clinical diagnostic criteria, the need for clear serum, saliva, or urine biomarkers to assess the physiologic burden of disease remains a work in progress, as are well-defined long-term clinically relevant outcomes of interest. Here, readers will be guided to a critical evaluation on the current main metrics of sleep study, suggested steps on diagnostic and treatment goals, and introduced to the concept of a continuum of disease.

In the light of the variability in the manifestation of obstructive sleep apnea (OSA) in patients and the knowledge gap regarding which patients will suffer cardiovascular consequences of this disease, significant controversy exists with respect to the selection of patients for OSA treatment and treatment outcomes. The proper evaluation and diagnosis of patients with OSA are critical for determining a suitable treatment plan. Clearly, a patient whose depression is misdiagnosed as OSA or, conversely, a patient whose OSA is misdiagnosed as depression will receive a treatment plan inappropriate for his or her clinical condition. In the evaluation of a patient with suspected sleep-disordered breathing, the history and physical examination, as well as various adjunctive studies and consultations, are helpful to correctly diagnose and treat the patient.

The current diagnostic methods for OSA include in-laboratory polysomnography and home portable monitoring, both of which report the apnea-hypopnea index (AHI) as the primary metric of OSA severity. Furthermore, the effectiveness of treatments for OSA is almost exclusively on reported changes in AHI; however, by focusing almost exclusively on the AHI the clinician will miss opportunities to better risk stratify patients by using other OSA-related variables.

Apnea-Hypopnea Index as the Single Gold Standard for Treatment Outcome?

In most, if not all, available studies on the health significance of OSA to date, the AHI has been used as the primary exposure or independent variable and has been correlated with specific outcomes of interest. The AHI is commonly used to determine OSA severity and, consequently, its association with cardiovascular disease [1, 2]. However, there is significant subjectivity, variability, and discrepancy in AHI scoring criteria.

Initially, it was recognized that apneic events fragmented sleep, induced cardiovascular instability, and led to excessive daytime sleepiness [3, 4]. Given the hemodynamic and sleep-related effects of obstructive apneas, it comes as no surprise that the apnea index, which tallies the number of apneas per hour of sleep, became the primary disease-defining metric for OSA. Over time, as the full spectrum of upper-airway collapse during sleep became more apparent, the simple concept of only quantifying apneas quickly evolved into something more complex [5]. It is now obvious that obstructive apneas, the original landmark for the disease, are not the only events of interest because obstructive hypopneas can also have similar effects (e.g., arousals and blood pressure swings) [5, 6].

Presently, the AHI depends largely on hypopnea scoring, which is a subject of great controversy, as mentioned in the latest Deliberations of the Sleep Apnea Definitions Task Force of the American Academy of Sleep Medicine report. Substantial disagreements remain regarding the degree of hypoventilation (or decrease in flow), severity of desaturation, and incorporation of arousals in defining a hypopnea [7, 8]. Also, several factors affect AHI scoring, including intraindividual (patient) AHI variability from night to night, variability between scorers, and lastly, sleep laboratories that vary in scoring criteria and/or interpretation of the rules. Although the argument that AHI is a valid measure of disease activity because it predicts the presence or development of clinical outcomes is reasonable, it is not without limitations and should be, at best, considered a crude and imprecise metric of OSA.

Oxygen Desaturation Index as the Valid Alternative?

The limitations of the AHI as a marker of physiologic information definitely led to the exploration of better metrics to correlate physiologic and clinical outcomes. A high-resolution oximeter is a watch oximeter with a high sampling frequency and resolution, which requires little training to install properly. It can detect the fluctuations in oxygen saturation caused by episodes of apneas and hypopneas. In addition, the data can be automatically analyzed with commercially available computer programs with an acceptable accuracy.

The oxygen desaturation index (ODI) is the number of times per hour of sleep that the blood oxygen level drops by 3 or 4% or more from baseline. As the improvement of cardiovascular disease risk is considered one of the purposes of treatment for OSA, the ODI may provide an alternative metric for the appraisal of treatment outcome. Furthermore, Punjabi et al. [2] demonstrated that hypopneas associated with desaturations are independently associated with cardiovascular disease; however, no association was observed between cardiovascular disease and hypopneas associated with arousals (no desaturations). Kendzerska et al. [9] demonstrated that other polysomnography outcomes, such as time with oxygen saturation below 90%, degree of sleep fragmentation, and total sleep time, were associated with worse cardiovascular outcomes, whereas AHI was not. Using nocturnal oximetry

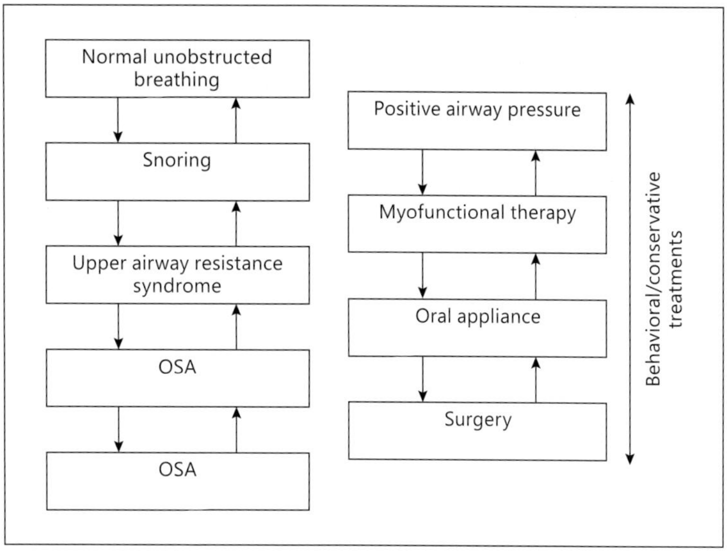

Fig. 1. The continuum of treatment and continuum of disease in obstructive sleep apnea (OSA).

to detect patients with sleep-disordered breathing or to appraise potential treatment outcomes is not a new idea, and it is a particularly attractive option because it can be easily generated with commercially available programs, and would allow the analysis of multiple nights. The clinical usefulness of ODI for the diagnosis of OSA still needs further clarification as it suffers from wide variations in terms of diagnostic parameters ranging from sensitivities of 31–98% and specificities of 41–100%, depending on the specific devices and ODI cutoffs used. However, the gold-standard of comparison in these approaches is the AHI and, as discussed previously, due to the significant influence of the definition of hypopnea, method of measurement, and interrater variability, the AHI has severe limitations in itself.

Clinically Relevant Outcomes, Treatment Options, and the Concept of Continuum of Treatment

It is now accepted that OSA is associated with major cardiovascular and noncardiovascular endpoints. Given the wealth of accumulated evidence after the initial clinical description of OSA, a robust disease-defining measure necessitates a better delineation of measures that not only depict the full spectrum of upper-airway collapse during sleep, but also accurately predict the occurrence of adverse events, and embrace the complexity of OSA with better measures. OSA is considered a chronic, presumably life-long condition, and the need to adjust, add or modify therapies is inescapable.

Furthermore, the real goal of treatment should not be solely to cure OSA in terms of AHI reduction, but most importantly to control symptoms and minimize ongoing multisystem damage. It is important to include both subjective (sleepiness, fatigue, unrefreshing sleep) and other objective, clinically relevant signs of the disease burden. It is likely that the definition of OSA may need to be revised to reflect not simply the AHI but the actual physiologic (degree and length of oxygen desaturation, heart rate variability changes, sleep fragmentation, sleep deprivation, and sympathetic activation) and patient-centered consequences (quality of life, patient and partner perception of disease and mortality) that occur.

In adults, nasal continuous positive airway pressure (CPAP) therapy is the gold standard treatment for OSA. Unfortunately, the long-term acceptance rate of CPAP therapy is around 70% in the best-designed trials. However, it is particularly low among younger patients, in unstructured clinical settings, and among those whose subjective ailments improve less with CPAP therapy. Consequently, many patients with moderate or severe OSA are referred to a secondary therapy (surgery, oral appliances, positional or myofunctional therapy).

A misconception in daily practice is that surgical and nonsurgical options are opposing treatments. In the literature, the concept of a continuum of disease is used frequently. The level of disease in an individual patient can improve or worsen, with the patient moving up or down this continuum both over time and on a night-to-night basis. To encompass the same complexity, OSA treatment can be viewed as a continuum from CPAP to surgical treatment or oral appliances at different time points during the disease or different patient expectations (Fig. 1). None of the treatments exclude the previous or the future treatment, and the continuum of treatment adapts to the continuum of disease. Parameters such as weight loss, patient preferences, compliance to treatment, and efficacy could move the therapy up or down this new continuum, requiring long-term monitoring by professionals from the various specialties involved in the treatment of OSA. In addition, in the case of nasal surgery, these parameters may complement treatment compliance to both CPAP and oral appliances. Moreover, behavioral/conservative treatments, such as weight reduction, optimizing of sleeping hygiene, avoidance of certain sleep positions, and drug treatment, can work synergistically with these other options.

References

1 Shahar E, Whitney CW, Redline S, et al: Sleep-disordered breathing and cardiovascular disease: cross-sectional results of the Sleep Heart Health Study. Am J Respir Crit Care Med 2001;163:19–25.

2 Punjabi NM, Newman AB, Young TB, Resnick HE, Sanders MH: Sleep disordered breathing and cardiovascular disease: an outcome-based definition of hypopneas. Am J Respir Crit Care Med 2008;177:1150–1155.

3 Lugaresi E, Coccagna G, Mantovani M, Cirignotta F, Ambrosetto G, Baturic P: Hypersomnia with periodic breathing: periodic apneas and alveolar hypoventilation during sleep. Bull Physiopathol Respir (Nancy) 1972;8:1103–1113.

4 Guilleminault C, Tilkian A, Dement WC: The sleep apnea syndromes. Annu Rev Med 1976;27:465–484.

5 Hudgel DW: "Apnea index": need for improving the description of respiratory variability during sleep. Am Rev Respir Dis 1986;133:708–709.

6 Gould GA, Whyte KF, Rhind GB, et al: The sleep hypopnea syndrome. Am Rev Respir Dis 1988;137:895–898.

7 Moser NJ, Phillips BA, Berry DT, Harbison L: What is hypopnea, anyway? Chest 1994;105:426–428.

8 Redline S, Sanders M: Hypopnea, a floating metric: implications for prevalence, morbidity estimates, and case finding. Sleep 1997;20:1209–1217.

9 Kendzerska T, Gershon AS, Hawker G, Leung RS, Tomlinson G: Obstructive sleep apnea and risk of cardiovascular events and all-cause mortality: a decade-long historical cohort study. PLoS Med 2014;11:1001599.

Prof. Robson Capasso
Department of Otolaryngology-Head and Neck Surgery
Division of Sleep Surgery
Stanford University School of Medicine
801 Welch Road
Stanford, CA 94305 (USA)
E-Mail rcapasso@stanford.edu

Lin H-C (ed): Sleep-Related Breathing Disorders. Adv Otorhinolaryngol. Basel, Karger, 2017, vol 80, pp 41–48
DOI: 10.1159/000470859

Updated Friedman Staging System for Obstructive Sleep Apnea

Michael Friedman[a, b] • Anna M. Salapatas[b] • Lauren B. Bonzelaar[c]

[a]Section of Sleep Surgery, Rush University Medical Center, and [b]Section of Otolaryngology, Chicago ENT, Advanced Center for Specialty Care, Advocate Illinois Masonic Medical Center, Chicago, IL, and [c]Department of Anesthesiology, Loyola University Medical Center, Maywood, IL, USA

Abstract

In the practice of sleep medicine, the first step is identification of those patients at high risk for sleep apnea. Nearly every physician and every hospital has preferred methods of screening. Many patient questionnaires or surveys as well as some objective physical measurements have been suggested to predict the presence of sleep apnea. Screening is well established, and laboratory and home testing are widely available. An early assessment with a physical examination can help direct treatment planning. The Friedman tongue position, lingual tonsil hypertrophy grading, and the effects of oral positioning on the hypopharynx should be used in early assessment for treatment planning, and as screening tools to assess the sight of obstruction. Although these screening tools are not substitutes for drug-induced sleep endoscopy (DISE), they are crucial in early assessment as many patients do not require surgery or DISE early in the evaluation.

Since the first description of uvulopalatopharyngoplasty (UPPP) by Fujita et al. [1] in 1981, surgical treatment of obstructive sleep apnea (OSA) has continued to evolve, with new techniques and modifications showing improvements in success. However, the efficacy of surgery for the treatment of sleep apnea remains controversial. Consideration of the complex relationship between different sites of obstruction is paramount to choosing the correct surgical approach to treatment. As such, the preoperative selection of patients that are likely to respond to surgery is both a necessity and priority. This requires the detailed assessment of individual patient anatomy. Correctly choosing patients who are likely to benefit from surgical treatment not only results in better outcomes, but also clarifies the need for surgery in specific populations of OSA patients.

OSA typically results from anatomic multilevel obstruction. There are many methods by which to determine sites of obstruction in OSA patients. Imaging (CT or MRI), cephalometric studies, drug-induced sleep endoscopy (DISE), and even snoring sound analysis are all methods that have been used to determine obstruction prior to surgery. It is important to have a strategy of airway evaluation that is both efficient and noninvasive for the early detection of surgical candidates. The performance of this type of outpatient screening is critical prior to consenting a patient for an invasive airway procedure. Based on simple clinical staging systems, patients are divided into 2 groups. Those with a favorable anatomy are directed to surgery after failing continuous positive airway pressure (CPAP). However, patients with an unfavorable anatomy are warned of the limited success of multilevel surgical procedures. If patients decide to proceed with surgery, DISE is a second preoperative assessment.

When evaluating a patient preoperatively, we use a well-established surgical staging system that incorporates the Friedman tongue position (FTP), tonsil size, and BMI. Surgical staging of OSA has been shown to be an accurate predictor of surgical success with UPPP, and can also be applied to predict patients who may be candidates for other treatments. Creating a universal language also provides insight on surgical success and morbidities that are likely to be associated with varying anatomic stages. We present our anatomic staging system and a grading system for lingual tonsil hypertrophy (LTH) that has recently been tested for interexaminer reliability.

Friedman Tongue Position

FTP is a simple means by which to approximate obstruction at the hypopharyngeal level. Descriptions of hypopharyngeal obstruction, prior to the introduction of FTP, were not standardized. Similar physical findings were reported with arbitrary terms such as "crowded oropharynx," "macroglossia," and "retrognathia," etc. This is confusing both in patient care and in reporting data.

FTP is the evaluation of the tongue's position relative to the tonsils/pillars, uvula, soft palate, and hard palate. It is based on Mallampati stages, which are used in the field of anesthesiology in the context of difficult endotracheal intubation [2]. While the Mallampati system has 3 grades and evaluates the palate with the tongue protruded, FTP evaluates the palate with the tongue in a neutral position inside the mouth. Furthermore, FTP is assigned on a scale from I to IV, and FTP II is split into IIa and IIb.

To evaluate FTP, the clinician should ask the patient to open his/her mouth widely, a minimum of 5 times. This allows the observer to assign the most consistent position of the tongue. The patient should also be instructed to "breathe normally," while leaving the tongue in its natural position. FTP I allows the observer to visualize the entire uvula and tonsils or pillars. FTP IIa allows visualization of the uvula but the tonsils are only partially seen. FTP IIb allows visualization of the complete soft palate down to the base of the uvula, but the uvula and the tonsils are not seen. FTP III allows visualization of some of the soft palate but the distal soft palate is eclipsed. FTP IV allows visualization of the hard palate only (Fig. 1).

Anatomic Grading System

The anatomic staging system integrates FTP, an estimation of hypopharyngeal obstruction, with tonsil size and BMI to predict surgical outcomes. This system can separate patients who will benefit from UPPP alone from those that will require multilevel surgical intervention. Tonsil size is graded from 0 to 4, where 0 is the absence of tonsillar tissue, size 1 represents tonsils contained within the pillars, size 2 tonsils extend to the pillars, size 3 tonsils extend beyond the pillars but not to the midline, and size 4 represents tonsils that extend to the midline (Fig. 2) [3, 4].

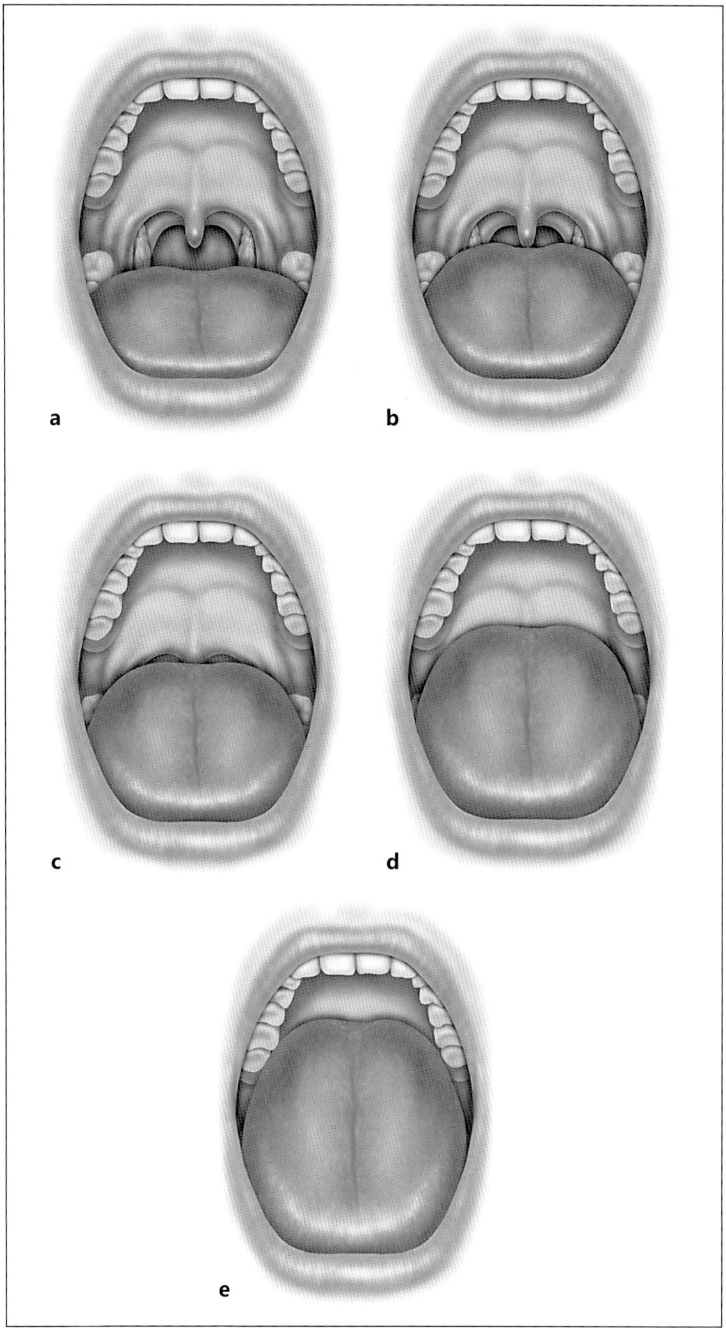

Fig. 1. Friedman tongue position (FTP). **a** FTP I visualizes the uvula and tonsils/pillar. **b** FTP IIa visualizes most of the uvula but not the tonsils/pillar. **c** FTP IIb visualizes the entire soft palate to the uvular base. **d** FTP III shows some of the soft palate with the distal end absent. **e** FTP IV visualizes only the hard palate. From Friedman [10].

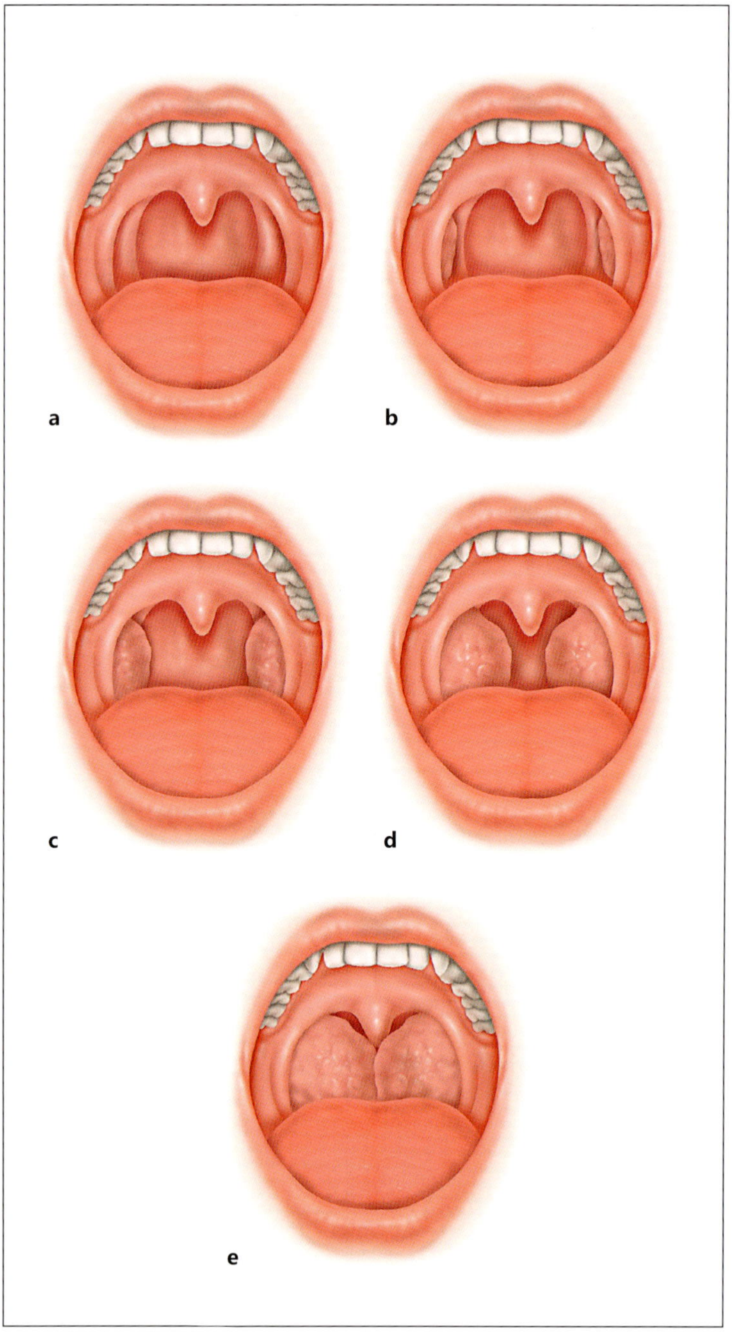

Fig. 2. Tonsil grading system. **a** Size 0, absence of tonsillar tissue. **b** Size 1, within the pillars. **c** Size 2, extended to the pillars. **d** Size 3, extended past the pillars. **e** Size 4, extended to the midline. From Friedman [10].

Table 1. Friedman staging system as determined by Friedman tongue position (FTP), tonsil size, and BMI

Stage	FTP	Tonsil size	BMI
I	I, IIa, IIb	3 or 4	<40
II	I, IIa, IIb III or IV	0, 1, or 2 3 or 4	<40 <40
III	III or IV	0, 1, or 2	<40
IV[a]	I–IV	0–4	>40

[a] All patients with significant craniofacial or other anatomic abnormalities.

Stage I disease is defined as patients with FTP I, IIa, or IIb, tonsil size 3 or 4, and BMI <40. Stage II disease is defined as FTP I, IIa, or IIb with tonsil size 0, 1, or 2, or FTP III or IV with tonsil size 3 or 4, and BMI <40. Stage III disease is defined as FTP III or IV, tonsil size 0, 1, or 2, and BMI <40. All patients with BMI >40 are classified as stage IV, alongside patients with skeletal deformities such as micrognathia or mid-face hypoplasia (Table 1).

Lingual Tonsil Hypertrophy Grading System

LTH is a common cause of hypopharyngeal obstruction. The degree of hypertrophy should be assessed in every patient as it reflects the degree of morbidity. Therefore, a universal language is necessary in order to compare the efficacy of removing different grades of lingual tonsils. The LTH grading system presented here may be useful in selecting patients requiring specific lingual tonsil surgery versus tongue base surgery.

To evaluate the lingual tonsils, the clinician should examine the awake patient with the tongue in multiple positions. The lingual tonsils are often best visualized with the tongue protruded, but this is not true in all patients. As multiple views of the tongue in different positions allows for a better understanding of the individual patient's LTH, awake endoscopy is superior to DISE for grading.

The grading system consists of a scale from 0 to 4. Grade 0 denotes a complete absence of lymphoid tissue on the tongue base. Grade 1 is defined as lymphoid tissue scattered over the tongue base. Grade 2 represents lymphoid tissue covering the entirety of the tongue base with limited vertical thickness. Grade 3 consists of significantly raised lymphoid tissue covering the entirety of the tongue base, approximately 5–10 mm in thickness. Grade 4 represents lymphoid tissue 1 cm or more in thickness, rising above the tip of the epiglottis (Fig. 3). The vertical depth of the lingual tonsils is a clinical approximation that should be judged by the otolaryngologist, with grade 3 being the first stage in which the tonsils have significant vertical height.

Discussion

In 1996, Sher et al. [5] reported that UPPP achieved surgical success in only 40% of patients in a meta-analysis of unselected patients. This brought forth not only the need for modifications of procedures to increase efficacy, but also the realization that not all OSA patients are suitable candidates for surgery. It was thought that surgery should be reserved for patients with mild/moderate disease, rather than severe disease. However, studies have shown that UPPP is not more effective in treating patients with mild disease, and can even worsen objective parameters such as the apnea-hypopnea index [6, 7].

The rationale for the anatomic staging system is that the success of UPPP is highly dependent on the anatomical relationship between palatal and hypopharyngeal obstruction. Patients who have large tonsils but minimal hypoglossal obstruction are likely to benefit from a procedure such as UPPP with tonsillectomy. Patients with

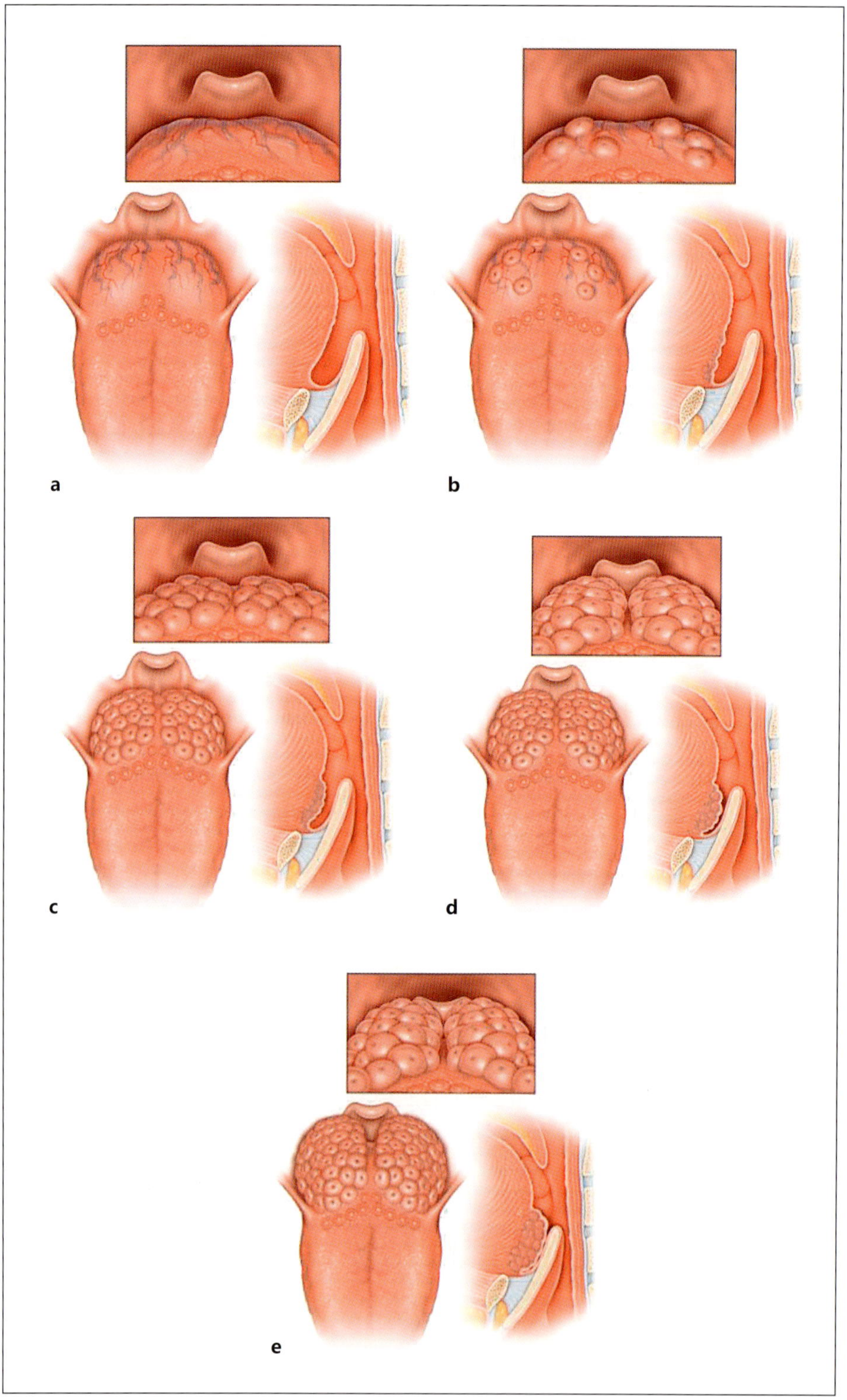

3

(For legend see next page.)

stage I disease have a greater than an 80% chance of success with UPPP, even if they have severe disease based on polysomnography. However, patients with stage III disease have more hypopharyngeal obstruction, and are unfavorable candidates for UPPP. Success rates for these patients, when treated with UPPP alone, are as low as 8% [8, 9] (Table 2). Treatment options for these patients would need to include plans for multilevel surgery. It is important to keep those with anatomic stage IV in mind; they are not surgical candidates and thus would also not benefit from DISE.

Among the developments in surgical treatment of OSA has been the introduction of transoral robotic surgery (TORS). It has become increasingly common to perform lingual tonsillectomy along with other TORS procedures. No studies have proven the value of lingual tonsillectomy for OSA. However, not all OSA patients have hypertrophied lingual tonsils. It is necessary to have a unified and cost-effective method of identifying patients who would benefit from lingual tonsillectomy and clarify which LTH grades are associated with perioperative and postoperative morbidity.

Table 2. Success and failure rates of uvulopalatopharyngoplasty (*n* with percentages in parentheses) for the treatment of obstructive sleep apnea stratified by stage

Stage	Unsuccessful	Successful	Total
I	6 (19.4)	25 (80.6)	31 (100)
II	18 (62.1)	11 (37.9)	29 (100)
III	68 (91.9)	6 (8.1)	74 (100)

Fig. 3. Lingual tonsil hypertrophy (LTH) grading system. **a** LTH 0, no lymphoid tissue. **b** LTH 1, scattered lymphoid tissue. **c** LTH 2, lymphoid tissue covering the entire tongue base, limited vertical thickness. **d** LTH 3, lymphoid tissue covering the entire tongue base, significant vertical thickness of approximately 5–10 mm. **e** LTH 4, lymphoid tissue covering the entire tongue base, rising to or above the tip of the epiglottis, approximately 1 cm in height. From Friedman et al. [11].

Generally, anatomic staging is used for the triage of patients who are noncompliant or intolerant to CPAP therapy. Grading of LTH is done to see whether specific patients would benefit from lingual tonsillectomy. It is our belief that LTH grade 3 and 4 are clinically significant and may be indicators for surgical removal of tonsillar tissue. To date, there are no studies that have identified the specific benefit of lingual tonsillectomy in relation to the degree of hypertrophy, especially since it is often done as a combined or adjunctive procedure with glossectomy. Similarly, no studies have identified the morbidity or mortality of lingual tonsillectomy as an isolated procedure. Clearly, however, there must be morbidity associated with the procedure. If we do perform lingual tonsillectomy, we can study both the efficacy and morbidity if we have a common language to describe lingual tonsil size. While logic would dictate that lingual tonsillectomy is beneficial for patients with LTH grades 3 and 4, it is not clear that the morbidity of lingual tonsillectomy is justifiable for LTH grades 1 or 2. Thus, a grading system is necessary to compare the results of excision based on the level of hypertrophy.

As it is an efficient tool to use in an outpatient setting, staging provides the clinician with valuable initial information about a patient's prospects of being a good surgical candidate. There are certainly further methods to directly visualize the patient's airway for areas of obstruction that should be utilized prior to operating. Nevertheless, anatomic staging remains a highly useful and noninvasive tool in patient selection.

Conclusion

The results of surgical treatment are never completely predictable for any disorder, but clinical staging offers important benefits. Staging systems are created to identify clinical features of the disease process that can predict whether any particular treatment option will be of value. The use of the anatomic staging system in OSA is a cost-effective, noninvasive, reproducible method to stratify patients based on anatomic variations. This can help improve surgical success rates and outcomes in OSA.

References

1 Fujita S, Conway W, Zorick F, Roth T: Surgical correction of anatomic abnormalities in obstructive sleep apnea syndrome: uvulopalatopharyngoplasty. Otolaryngol Head Neck Surg 1981;89: 923–934.

2 Mallampati SR, Gatt SP, Gugino LD, Desai SP, Waraksa B, Freiberger D, Liu PL: A clinical sign to predict difficult tracheal intubation: a prospective study. Can Anaesth Soc J 1985;32:429–434.

3 Friedman M, Tanyeri H, La Rosa M, Landsberg R, Vaidyanathan K, Pieri S, Caldarelli D: Clinical predictors of obstructive sleep apnea. Laryngoscope 1999;109:1901–1907.

4 Friedman M, Ibrahim H, Bass L: Clinical staging for sleep-disordered breathing. Otolaryngol Head Neck Surg 2002;127: 13–21.

5 Sher AE, Schechtman KB, Piccirillo JF: The efficacy of surgical modifications of the upper airway in adults with obstructive sleep apnea syndrome. Sleep 1996; 19:156–177.

6 Friedman M, Vidyasagar R, Bliznikas D, Joseph N: Does severity of obstructive sleep apnea/hypopnea syndrome predict uvulopalatopharyngoplasty outcome? Laryngoscope 2005;115:2109–2113.

7 Senior BA, Rosenthal L, Lumley A, Gerhardstein R, Day R: Efficacy of uvulopalatopharyngoplasty in unselected patients with mild obstructive sleep apnea. Otolaryngol Head Neck Surg 2000;123: 179–182.

8 Friedman M, Ibrahim H, Joseph NJ: Staging of obstructive sleep apnea/hypopnea syndrome: a guide to appropriate treatment. Laryngoscope 2004;114: 454–459.

9 Li H, Wang P, Lee L, Chen N, Fang T: Prediction of uvulopalatopharyngoplasty outcome: anatomy-based staging system versus severity-based staging system. Sleep 2006;29:1537–1541.

10 Friedman M: Friedman tongue position and the staging of obstructive sleep apnea/hypopnea syndrome; in Friedman M (ed): Sleep Apnea and Snoring: Surgical and Non-Surgical Therapy. Philadelphia, Saunders Elsevier, 2009, chapt 16.

11 Friedman M, Yalamanchali S, Gorelick G, Joseph NJ, Hwang MS: A standardized lingual tonsil grading system: interexaminer agreement. Otolaryngol Head Neck Surg 2015;152:667–672.

Michael Friedman, MD
ChicagoENT
30 North Michigan Avenue, Suite 1107
Chicago, IL 60602 (USA)
E-Mail mfriedman@chicagoent.com

Lin H-C (ed): Sleep-Related Breathing Disorders. Adv Otorhinolaryngol. Basel, Karger, 2017, vol 80, pp 49–56
DOI: 10.1159/000470895

Advancements of CPAP Therapy for Obstructive Sleep Apnea

Ning-Hung Chen

Sleep Center, Department of Pulmonary and Critical Care Medicine, Chang Gung Memorial Hospital Linkou, and Respiratory Therapy, Chang Gung University, Taoyuan City, Taiwan, ROC

Abstract

Continuous positive airway pressure (CPAP), the gold standard for the treatment of patients with obstructive sleep apnea (OSA) syndrome, was introduced by Sullivan in 1981. CPAP acts as an air splinter to support the collapsing upper airway during sleep. The mask, the interface between CPAP and the patient, plays a major role in patient acceptance and compliance with treatment. Improvements in mask materials, design, and aerodynamics have over time continued to increase patient compliance in the use of CPAP. Optimized pressure titration and mask fitting, as well as educational programs, are key factors that have increased the likelihood of CPAP acceptance, as well as teamwork between experienced technicians and enthusiastic physicians. The numerous potential side effects of CPAP, which include nasal and flow-related problems, skin abrasion, and humidifier issues, can be resolved with appropriate and frequent supervision during the early phase of treatment and including regular long-term reviews. CPAP can improve the cardiovascular, neurocognitive, and renal hepatic consequences of OSA, thus improving longevity and quality of life as well as reducing daytime somnolence. When compared with the recent technique of auto-titrating CPAP (APAP), traditional fixed-pressure CPAP has also been proven to be well accepted in appropriately selected patients. Patient compliance and improvement in daytime sleepiness favor APAP, but fixed-pressure CPAP provides a higher oxygen saturation during sleep. In conclusion, CPAP is the most accepted treatment modality for OSA, but to achieve higher patient acceptance rates and better compliance, research needs to continue into the facial apparatus and delivery methods of the oxygen.

Continuous positive airway pressure (CPAP) is the gold standard for the treatment of patients with obstructive sleep apnea (OSA) syndrome. Other treatment modalities for OSA have also been advocated, including surgery such as uvulopalatoplasty and mandibular advancement. CPAP remains the treatment of choice for the majority of patients, with a large volume of research evidence supporting positive outcomes.

The Development of Continuous Positive Airway Pressure, 1981–1985

Nasal CPAP was described by Sullivan et al. [1] 1981. They successfully treated 5 patients with severe OSA using CPAP through the nares via a comfortable nose mask. CPAP provides a pneumatic splint for the nasopharyngeal airway [1]. Other treatments were introduced around the same time, such as tracheostomy and uvulopalatal surgery, thus introducing clinician bias regarding the best option for patient treatment. By 1985, CPAP was recognized as the treatment most likely to be effective for OSA. This nonsurgical method of treating OSA was commercialized by the ResMed Company, and their first CPAP machine was named after Dr. Sullivan.

These first "fixed-pressure" CPAP machines had some initial drawbacks. Patients complained of the machine noise and the local facial discomfort from its pressure delivery system, as well as its poor interface, which made the regular use of CPAP difficult. Furthermore, the delivery of pressure to open the upper airway changed frequently depending on patient position and stage of sleep, which resulted in patients being roused during sleep due to pressure overshooting. To combat this sleep interference due to body posture, the introduction of auto-titrating CPAP (APAP) was developed and became available around the year 2000.

Mode of Continuous Positive Airway Pressure: Balance of Pressure

As described by Sullivan et al. [1], CPAP acts as an air splinter to open the collapsing upper airway during sleep. The mode of action follows the "tube rule." The upper airway is considered to be a tube and the pressure across the tube should be balanced in order to keep the tube open. The outer pressure of the tube consists of upper airway tissue (tissue pressure, Pti) and atmospheric pressure; the inner pressure consists of the pressure generated from airflow (luminal pressure, Pl) and the pressure from the atmosphere. When the muscle tone of the upper airway is quiet, the upper airway follows the tube law. That is, if Pl is higher than Pti, the airway will remain open. When Pti is higher than Pl, the airway closes. When inspiration occurs during sleep, the intraluminal pressure becomes negative to suck air into the lungs. That renders the Pl negative, facilitating the possible collapsing of the upper airway. The other force acting on the upper airway that is necessary to keep it open is the neuromuscular activity of the upper airway muscles.

CPAP acts to increase the intraluminal airway pressure, Pl, to counteract both the pressure generated from the tissues and the negative intraluminal suction pressure. As the intraluminal pressure of inspiration changes during different stages of respiration and the neuromuscular activity also changes during different sleep stages, the optimal pressure needed to keep the airway open changes dynamically [2].

The Application of Continuous Positive Airway Pressure

Experienced technicians and enthusiastic physicians are the major factors in ensuring the successful application of CPAP.

Mask Fitting

The mask delivery to the airway is the most important factor that determines patient acceptance of CPAP. A full facemask, nasal mask, and nasal prong or pillow are the most widely available appliances for the delivery of CPAP. A nasal prong or pillow was initially recommended for patients who needed low-pressure oxygen, slept on their sides, were claustrophobic, or had a proven skin allergy to silicon. However, increasing numbers of patients favor this small interface method, which makes the initial use of CPAP more acceptable. However, the small contact area increases the possibility of an air leak.

Full facial masks have been preferred in many countries as this traditional interface can prevent air leakage from patients who breathe through open mouths. However, some studies have shown that pressure may increase during use (compared to that which occurs when a nasal mask is used) due to the interaction of airflows from both the nose and mouth.

The nasal mask has also found preference in certain countries because it is large enough to remain stable but is not as large as the full facemask, making it more acceptable to patients [3]. An informal survey conducted by a CPAP company found a trend of full facemask use in North America and Europe, and nasal mask use in Asian countries, although the reason was not well elucidated.

A well-designed mask has become the major challenge for CPAP companies. Current masks are designed for mass production and may fit the majority of users well. However, pediatric OSA patients, craniofacial anomalies, different facial profiles between ethnicities, or extreme phenotype faces all represent cases in which it can be difficult to find a suitable mask. Tailor-made masks that fit well should be a future goal of mask development.

CPAP Titration and Pressure Setting

The optimal pressure for a patient should be determined after overnight CPAP titration. Although APAP is widely used in many countries, overnight titration is highly recommended, especially in patients who cannot tolerate APAP. The overnight titration protocol recommended by the AASM is summarized as follows. After the patient falls asleep with APAP, the pressure should be elevated by 1 cm H_2O whenever 2 episodes of obstructive apnea, 3 episodes of hypopnea, 5 respiratory effort-related arousals, or 3 min of loud snoring occur during a 5-min period. Bilevel-PAP should be considered when the pressure is elevated to 15 cm H_2O. If patients have no breathing event during 30 min of normal sleep, or during 15 min of supine REM, one could attempt to lower the pressure by 1 cm H_2O every 10 min. Optimal pressure should be obtained for patients without any respiratory events (including loud snoring) in every position and sleep stage.

Netzer et al. [4] suggested that pressure will change by up to 2 cm H_2O after 2 months of CPAP use. Retitration of the pressure may be needed if patients are unable to tolerate the initial pressure settings. Ramp time settings developed according to the sleep latency of the patient may also be needed when using fixed pressure CPAP. This ramp setting is designed to make the beginning of therapy more comfortable. It begins by delivering a low initial pressure to help the patient to fall asleep more easily.

The setting of the pressure range is complicated as no recommended standard pressure has been advocated. Some sleep labs suggest 4 cm H_2O above and below the titrated optimal pressure; however, a default pressure of 4–20 cm H_2O in the APAP machine has also been recommended at the beginning of prescription. Adjustment of the APAP pressure may be needed after several months of use if a patient has complaints. With the development of the technique for auto CPAP and telemedicine, increasingly more sleep labs have their own titration protocols. Some are developing home titration with acceptable results, although further studies with better evidence are needed for such home titration protocols.

Humidifier and/or Heated Tube

Humidifiers had to be purchased separately for early machines, thus some patients frequently experienced dryness and nasal complaints after CPAP. Humidifier use reduces these problems, although they remain frequently noted. The humidifier power depends on the brand make of the machine, meaning patients experience different humidifier issues, including dry throat, water condensation in the mask or tube, and nasal discharge. Patient education regarding the adjustment of the humidifier is necessary before CPAP use. Figure 1 tabulates patient complaints at the

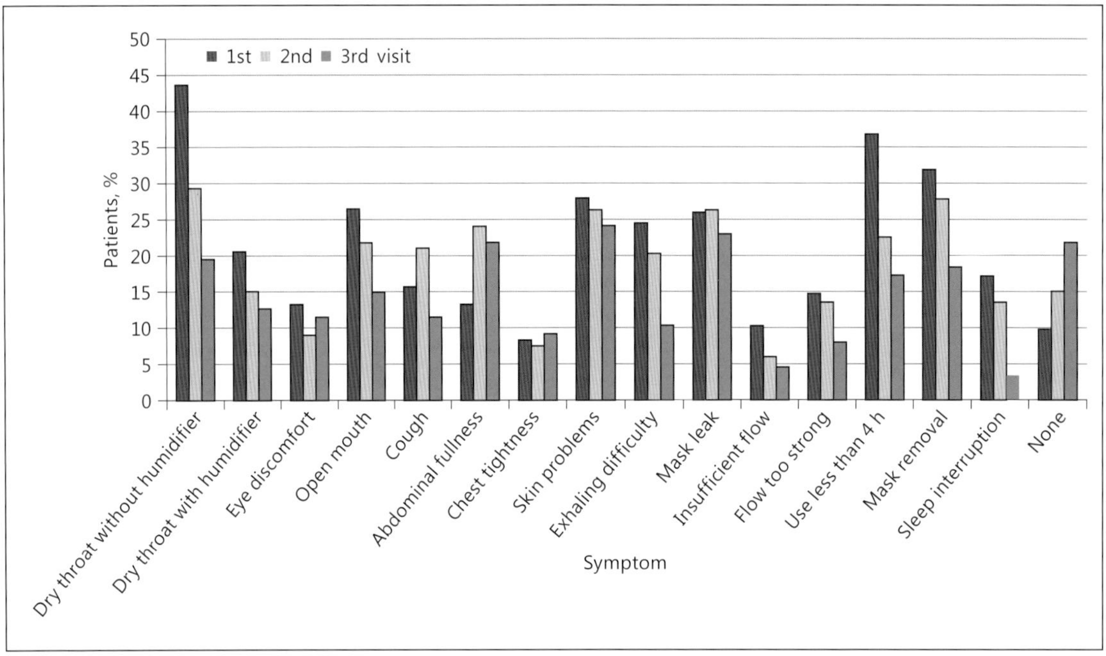

Fig. 1. Discomfort reported by 200 obstructive sleep apnea patients using CPAP following their first 3 follow-up clinic visits.

author's sleep center after CPAP usage resulting from 3 sequential CPAP clinics.

Side Effects of Continuous Positive Airway Pressure

CPAP is relatively safe for patients with OSA. However, some discomfort and rare complications have been noted following long-term use. The more common side effects include nasal, mask, and flow-related problems. The most frequent complaints and the related methods to solve these problems are summarized in Table 1. Figure 2 also shows the frequent side effect of nasal skin abrasion from mask use.

Some Dangerous, but Rare, Side Effects
Dangerous side effects from CPAP are rare. However, patients with altered consciousness or inadequate muscle power to remove the mask are relative contraindications for CPAP use. Warnings of aspiration, suffocation, or sputum impaction should be given to the caregivers of these patients. Pulmonary barotrauma, such as pneumothorax, pneumomediastinum, subcutaneous emphysema, or pneumocephalus, after CPAP use have been reported in patients with facial trauma or head injury. Increased intraocular pressure and tympanic membrane rupture have also been reported in some cases, as has periodontitis.

Evidence for Continuous Positive Airway Pressure Treatment

CPAP as a treatment for OSA has much more positive and supporting evidence of a satisfactory patient outcome compared with other treatment

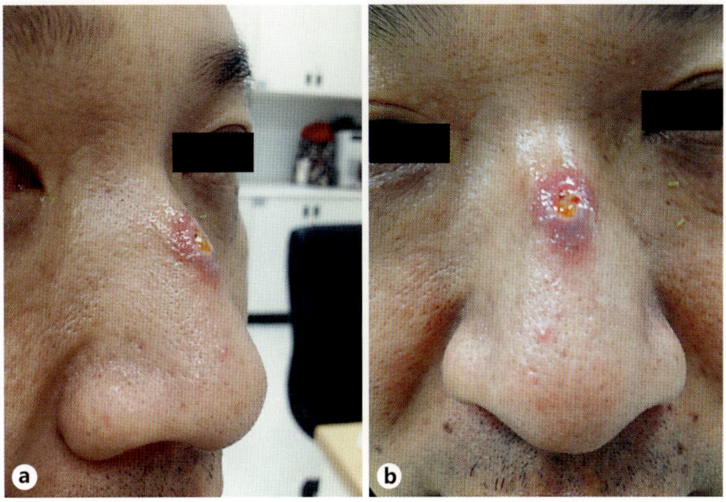

Fig. 2. A typical side effect of nasal mask use: skin abrasion after 1 month of CPAP treatment.

Table 1. Side effects of CPAP

Problem	Complaints	Solution
Nose	Rhinorrhea Nasal congestion Oronasal dryness Epistaxis	Humidifier Nasal decongestants/nasal steroids ENT doctor consultation
Mask and skin	Skin abrasion and wounds Skin itching, irritation, or rash Conjunctivitis Corneal injury Claustrophobia	Adequate education Fitted mask and headgear Mask cleansing Reassurance of safety
Flow	Chest discomfort or pain Aerophagia with abdominal fullness or flatus passage Sinus discomfort Difficulty exhaling	Adequate pressure setting (retitration) Pressure relief technique APAP instead of CPAP

modalities. The likely treatment outcomes from CPAP, including different anatomic systems and patient well-being, are discussed below.

Cardiovascular Effects

There is much evidence to show that CPAP can reduce cardiovascular complications in patients with OSA. In 2005, Marin et al. [5] reported that patients with OSA maintained the same level of cardiovascular outcome after CPAP treatment over 12 years of observation. Peker et al. [6] also reported an increased risk of cardiovascular disease, including hypertension, coronary artery disease, and cardiovascular events in OSA patients, and these risks returned to normal after CPAP treatment for 7 years. Doherty et al. [7] reported a significant decrease in survival after 25 months in 168 patients with an apnea-hypopnea index of more than 15 who were not subjected to CPAP treatment. Furthermore, the Spanish Sleep Network also reported a 3.1 mm Hg decrease in 24-h mean blood pressure and a 3.2 mm Hg decrease in 24-h diastolic blood pressure (DBP) after CPAP [8]. A significant positive correlation was also reported between hours of CPAP and decrease in 24-h mean blood pressure [9]. A meta-analysis [10] of 16 randomized controlled trials in 1,162 patients showed that the use of CPAP resulted in reductions in clinic-measured systolic blood pressure (SBP; –3.20 mm Hg, 95% CI –4.67 to –1.72) and DBP (–2.87 mm Hg, 95% CI –5.18 to –0.55), in nighttime SBP (–4.92 mm Hg, 95% CI –8.70 to –1.14), in mean 24-h blood pressure (–3.56 mm Hg, 95% CI –6.79 to –0.33), in mean nighttime blood pressure (–2.56 mm Hg, 95% CI –4.43 to –0.68), and in 24-h DBP (–3.46 mm Hg, 95% CI –6.75 to –0.17). However, no significant change was observed in daytime SBP (–0.74 mm Hg, 95% CI –3.90 to 2.41) or daytime DBP (–1.86 mm Hg, 95% CI –4.55 to 0.83).

Patel et al. [11] reported a significantly higher success rate of catheter ablation of nonparoxysmal atrial fibrillation in patients with OSA when they used CPAP (79% in the CPAP group vs. 68% in the non-CPAP group, $p = 0.032$). Fein et al. [12] also reported a significantly greater arrhythmia-free survival rate in OSA patients who were "CPAP users" when compared with "CPAP nonusers" after 1 year (71.9 vs. 36.7%, $p = 0.01$). Two review articles summarize the effect of CPAP on decreasing blood pressure and increasing left ventricular ejection fraction from 2 to 9% [13, 14].

Neurocognitive Effects

In all studies, CPAP has been shown to reduce the incidence of traffic accidents. A meta-analysis of 9 observational studies examining the crash risk of drivers with OSA before compared to after CPAP found a significant risk reduction after CPAP treatment (risk ratio 0.278, 95% CI 0.22–0.35; $p < 0.001$) [15]. Daytime sleepiness is the second most common complaint in patients with OSA. Sleepiness, as measured by subjective questionnaires such as the Epworth Sleepiness Scale or Stanford Sleepiness Scale, or by objective methods such as the Multiple Sleep Latency Test or Osler Test, were significantly improved after CPAP use, as shown in randomized controlled trials comparing improvements in sleepiness in patients using CPAP [16–19]. The largest benefit was in patients with the most severe sleepiness before treatment [20].

Neurocognitive deficits, other than motor performance and sleepiness, are variable in patients with OSA. Deficits in attention, memory, intelligence, and executive function have been reported. Results from studies concerning neurocognitive dysfunction are inconclusive, possibly due to small patient numbers or the study methods employed.

Quality of Life and Sleepiness

Quality of life in OSA patients is significantly impaired. Most studies using the Short Form 36 (SF-36) questionnaire reported an impairment in all domains, or at least the majority of them. Ambrosio et al. [21] reported impairment in all aspects of quality of life in OSA patients, as measured by SF-36, and nasal CPAP improved vitality, social functioning, and mental health. The Sleep Heart Health Cohort Study (6,440 men and women over 40 years of age) found an association between severe sleep-disordered breathing and the SF-36 quality of life scale, especially regarding physical function, general health, vitality, and social function. McArdle et al. [22] reported that the partners of OSA patients had a poor sleep quality and health status, and were shown to benefit from their partner's CPAP treatment. Two studies from Taiwan also showed an impairment of physical function, general health, social function, vitality, emotional function, and mental health (when compared to the general population), all of which improved after CPAP treatment [23, 24].

Other Physical Impairments

Renal dysfunction, liver dysfunction, and metabolic syndromes have been frequently reported. Normal tension glaucoma is another manifestation of OSA that is frequently cited.

Table 2. Comparison between APAP and fixed-pressure CPAP

Advantages of APAP	Equal	Advantages of CPAP
Compliance 0.18 h/night 95% CI 0.05 to 0.31 $p = 0.006$	Arousal index –0.85 95% CI –2.2 to 0.5 $p = 0.23$	Minimum O_2 –1.3% 95% CI –2.2 to –0.4 $p = 0.003$
Epworth Sleepiness Scale –0.48 95% CI –0.81 to –0.15 $p = 0.005$	Apnea-hypopnea index 0.25 95% CI –0.16 to 0.66 $p = 0.23$	

Fixed-Pressure (CPAP) versus Auto-Titration (APAP)

Studies have demonstrated CPAP pressure changes in the majority of patients several weeks after the commencement of therapy [4]. Randerath et al. [25] performed a randomized controlled trial of CPAP versus APAP and reported that, for all BMI values, a significant difference was seen between CPAP pressure and mean pressure using APAP. Using APAP, the pressure was lower than the pressure using CPAP for 81.5 ± 21% of the time [25]. Another study showed that the mean positive airway pressure was lower in APAP but had no influence on therapy adherence, clinical outcome, or compliance [26]. Since becoming available in 2000, algorithms have been developed for every type of APAP machine. Schwab et al. [27] summarized the events-detection principle in 2013. In summary, apnea was detected in different machines from a flow reduction of 75, 80, or 90% in 10 s. Hypopnea is detected by the flow falling from 40 to 80% depending on the algorithm.

Some machines can detect a limitation in flow, while others cannot. Therefore, the responses to sleep respiratory events vary, which may make patients more tolerant of certain types of machine over others. Table 2 summarizes the principle differences between APAP and CPAP. In a systemic review of CPAP versus APAP, Ip et al. [28] found that in individual studies APAP and CPAP resulted in similar changes in apnea-hypopnea index from baseline, as well as in most other sleep study measures, including quality of life. By meta-analysis, APAP improved compliance by 11 min/night (95% CI 3–19) and reduced sleepiness as measured by the Epworth Sleepiness Scale by 0.5 points (95% CI 0.8–0.2) compared with fixed CPAP. However, fixed CPAP improved minimum oxygen saturation by 1.3% more than APAP (95% CI 0.4–2.2).

In summary, APAP improves the compliance of patients at the expense of oxygenation during sleep. However, the treatment effects of either type of machine are similar. Specialist clinicians should consider each patient's needs individually when selecting the most suitable machine available before commencing treatment.

Conclusion

CPAP is the gold standard of treatment for the majority of OSA patients. CPAP has much more evidence to support its efficacy compared with other treatment modalities. Physicians should make every effort to improve CPAP acceptance and compliance, and recommend CPAP as the first choice of treatment for the majority of OSA patients.

References

1 Sullivan CE, Issa FG, Berthon-Jones M, et al: Reversal of obstructive sleep apnoea by continuous positive airway pressure applied through the nares. Lancet 1981;1:862–865.

2 Schwab RJ, Remmers JE, Kuna ST: Anatomy and physiology of upper airway obstruction; in Kryger MH, Roth T, Dement WC (eds): Principle and Practice of Sleep Medicine, ed 5. Amsterdam, Elsevier, 2011, chapt 101, pp 1155–1156.

3 Andrade RG, Piccin VS, Nascimento JA, et al: Impact of the type of mask on the effectiveness of and adherence to continuous positive airway pressure treatment for obstructive sleep apnea. J Bras Pneumol 2014;40:658–668.

4 Netzer NC, Juhász J, Hofmann M, et al: The need for pressure changes in CPAP therapy 2–3 months after initial treatment. Sleep Breath 2011;15:107–112.

5 Marin JM, Carrizo SJ, Vicente E, et al: Long-term cardiovascular outcomes in men with obstructive sleep apnoea-hypopnoea with or without treatment with continuous positive airway pressure: an observational study. Lancet 2005;365:1046–1053.

6 Peker Y, Hedner J, Norum J, et al: Increased incidence of cardiovascular disease in middle-aged men with obstructive sleep apnea: a seven-year follow-up. Am J Respir Crit Care 2002;166:159–165.

7 Doherty LS, Kiely JL, Swan V, et al: Long-term effects of nasal continuous positive airway pressure therapy on cardiovascular outcomes in sleep apnea syndrome. Chest 2005;127:2076–2084.

8 Campos-Rodriguez F, Perez-Ronchel J, Grilo-Reina A, et al: Long-term effect of continuous positive airway pressure on blood pressure in patients with hypertension and sleep apnea. Chest 2007;132:1847–1852.

9 Martínez-García MA, Capote F, Campos-Rodríguez F, et al; Spanish Sleep Network: Effect of CPAP on blood pressure in patients with obstructive sleep apnea and resistant hypertension: the HIPARCO randomized clinical trial. JAMA 2013;310:2407–2415.

10 Schein ASO, Kerkhoff AC, Coronela CC, et al: Continuous positive airway pressure reduces blood pressure in patients with obstructive sleep apnea: a systematic review and meta-analysis with 1,000 patients. J Hypertens 2014;32:1762–1773.

11 Patel D, Mohanty P, Di Biase L, et al: Safety and efficacy of pulmonary vein antral isolation in patients with obstructive sleep apnea: the impact of continuous positive airway pressure. Circ Arrhythm Electrophysiol 2010;3:445–451.

12 Fein AS, Shvilkin A, Shah D, et al: Treatment of obstructive sleep apnea reduces the risk of atrial fibrillation recurrence after catheter ablation. J Am Coll Cardiol 2013;62:300–305.

13 Bradley TD, Floras JS: Obstructive sleep apnoea and its cardiovascular consequences. Lancet 2009;373:82–93.

14 Haentjens P, van Meerhaeghe A, Moscariello A, et al: The impact of continuous positive airway pressure on blood pressure in patients with obstructive sleep apnea syndrome: evidence from a meta-analysis of placebo-controlled randomized trials. Arch Intern Med 2007;167:757–764.

15 Tregear S, Reston J, Schoelles K, et al: Continuous positive airway pressure reduces risk of motor vehicle crash among drivers with obstructive sleep apnea: systematic review and meta-analysis. Sleep 2010;33:1373–1380.

16 Crawford MR, Bartlett DJ, Coughlin SR, et al: The effect of continuous positive airway pressure usage on sleepiness in obstructive sleep apnoea: real effects or expectation of benefit? Thorax 2012;67:920–924.

17 Patel SR, White DP, Malhotra A, et al: Continuous positive airway pressure therapy for treating sleepiness in a diverse population with obstructive sleep apnea: results of a meta-analysis. Arch Intern Med 2003;163:565–571.

18 Marshall NS, Barnes M, Travier N, et al: Continuous positive airway pressure reduces daytime sleepiness in mild to moderate obstructive sleep apnoea: a meta-analysis. Thorax 2006;61:430–434.

19 McDaid C, Durée KH, Griffin SC, et al: A systematic review of continuous positive airway pressure for obstructive sleep apnoea-hypopnoea syndrome. Sleep Med Rev 2009;13:427–436.

20 Craig SE, Kohler M, Nicoll D, et al: Continuous positive airway pressure improves sleepiness but not calculated vascular risk in patients with minimally symptomatic obstructive sleep apnoea: the MOSAIC randomized controlled trial. Thorax 2012;67:1090–1096.

21 Ambrosio SD, Bowman T, Mohsenin V: Quality of life in patients with obstructive sleep apnea. Chest 1999;115;123–129.

22 McArdle N, Kingshott R, Engleman HM, et al: Partner of patients with sleep apnoea/hypopnoea syndrome: effect of CPAP treatment on sleep quality and quality of life. Thorax 2001;56:513–518.

23 Wang PC, Li HY, Shih TS, et al: Generic and specific quality-of-life measures in Taiwanese adults with sleep-disordered breathing. Otolaryngol Head Neck Surg 2006;135:421–426.

24 Lin SW, Chen NH, Li HY, et al: A comparison of the long-term effects of surgery or continuous positive airway pressure on the general health of patients with obstructive sleep apnea syndrome. Laryngoscope 2006;116:1012–1016.

25 Randerath WJ, Schraeder O, Galetke W, et al: Autoadjusting CPAP therapy based on impedance efficacy, compliance and acceptance. Am J Respir Crit Care Med 2001;163:652–657.

26 Damjanovic D, Fluck A, Bremer H, et al: Compliance in sleep apnoea therapy: influence of home care support and pressure mode. Eur Respir J 2009;33:804–811.

27 Schwab RJ, Badr SM, Epstein LJ, et al: An official American Thoracic Society statement: continuous positive airway pressure adherence tracking systems: the optimal monitoring strategies and outcome measures in adults. Am J Respir Crit Care Med 2013;188:613–620.

28 Ip S, D'Ambrosio C, Patel K, et al: Autotitrating versus fixed continuous positive airway pressure for the treatment of obstructive sleep apnea: a systematic review with meta-analyses. Syst Rev 2012;1:20.

Ning-Hung Chen
Sleep Center, Department of Pulmonary and Critical Care Medicine
Chang Gung Memorial Hospital, 5 Fuhsing Street
Guishan, Taoyuan 333, Taiwan (ROC)
E-Mail nhchen@adm.cgmh.org.tw

Lin H-C (ed): Sleep-Related Breathing Disorders. Adv Otorhinolaryngol. Basel, Karger, 2017, vol 80, pp 57–65
DOI: 10.1159/000470865

Advances in Oral Appliances for Obstructive Sleep Apnea

Ofer Jacobowitz

ENT and Allergy Associates, New York, NY, USA

Abstract

Oral appliances that advance the mandible are widely used as alternatives to positive airway pressure (PAP) devices or as primary therapy for obstructive sleep apnea (OSA) in adults. Although PAP is more efficacious at lowering the polysomnographic indices of OSA, the clinical effectiveness of PAP and oral appliances is similar, and patients are more likely to adhere to oral appliance therapy than to PAP treatment. Clinical examination is used to determine the candidacy of oral appliances and to select a particular appliance for a given patient. Endoscopic examination of the pharynx may be used to help assess the potential for efficacy. Otherwise, if available, titration of mandibular protrusion during sleep may be performed prior to appliance production in order to assess efficacy. Once a patient is fitted with a titratable oral appliance, further advancement is usually performed at home to resolve the clinical symptoms and signs of OSA. Clinical follow-up is needed to assess the outcome, side effects, and adherence, as the long-term adherence rate is approximately 50%. Recent advances in oral appliance therapy include the development of embedded temperature sensors for adherence monitoring and the production of thinner, lighter appliances via 3D printing techniques.

Obstructive sleep apnea (OSA) is a highly prevalent medical disorder that is challenging to treat. The consequences of untreated sleep apnea and associated sleep disruption include cardiovascular, neurological, metabolic dysfunction, poor functional performance, social isolation, motor vehicle accidents, and premature death. OSA occurs in sleep due to a reduction of upper airway muscle tone within a structurally small pharyngeal airway, but modifying factors include the variability of muscle tone, arousal threshold, respiratory control, upper airway inflammation, nasal obstruction, and head position. Positive airway pressure (PAP) is the most commonly prescribed treatment for OSA, and restores upper airway patency as a pneumatic splint. While PAP is efficacious at normalizing the apnea-hypopnea index (AHI), it is often not accepted or is not used in the long term. PAP adherence rates are low, and may be <50%, even when studied in a clinical trial setting where adherence would be expected to be higher [1]. Alternative treatments are thus in great need and include upper airway reconstructive surgery and oral appliance therapy (OAT).

OAT is an accepted and validated treatment modality for OSA in adults [2]. The most common and best studied type of oral appliances for OSA are those that advance the mandible forward, often called mandibular repositioning devices (MRDs). The main mechanism of action of MRDs is stabilization and dilation of the upper airway, in particular at the level of the velopharynx. Fiber-optic endoscopic examination of the pharynx during mandibular protrusion can be used to help predict the potential efficacy of OAT. Most often, however, only empiric therapy is used with subsequent sleep study assessment. Adherence to OAT is dependent on comfort, nasal patency, effectiveness, and the presence of side effects, but is usually greater than adherence to continuous PAP (CPAP). The side effects of OAT include oral or jaw discomfort, occlusal changes, and dry mouth, and are the most common reason for treatment discontinuation. Nonetheless, patients usually prefer oral appliances to PAP therapy and are about twice less likely to discontinue OAT as compared with PAP [2].

Efficacy and Effectiveness

Efficacy of a treatment for OSA is defined as its ability to treat upper airway obstruction in the ideal setting, most often measured in terms of AHI reduction to a given threshold. Effectiveness, however, takes into account actual adherence to the treatment in the home setting, and the ability of a treatment to ameliorate the signs and symptoms of OSA. For surgical treatment, efficacy equals effectiveness as adherence is 100%. PAP therapy almost always normalizes the AHI during titration in a sleep center, but in the home setting the AHI may not be normal even at an apparently efficacious setting. In 1 study, while on PAP therapy, AHI <5 was only present in 27% of patients and AHI <10 was present in 61% of patients followed in highly skilled sleep medicine programs [3]. Another factor for effectiveness is the AHI in the home setting when adjusted to the number of hours used at night [4]. A patient with severe OSA who only uses PAP for 4 out of 7 h per night may be considered to only be 57% treated in terms of the all-night AHI, or otherwise as treated for 4 h and untreated with severe OSA for 3 h. Furthermore, long-term adherence to PAP is probably close to 50%, and in a recent large-scale study, adherence to PAP for 4 h or more per night was confirmed as 42% at 3 years from the onset of therapy [5]. Thus, alternative treatments that either do not depend on adherence, such as surgery, are needed, or otherwise those that have higher adherence rates, such as OAT.

OAT efficacy depends on the degree of mandibular protrusion and the consequent stabilization and dilation of the upper airway. Efficacy is thus much more variable than that of PAP, which as a pneumatic splint can dilate any airway that does not have a rigid obstruction. PAP is more efficacious than MRDs at reducing the AHI by about 6 events per hour to <5 [2]. For MRDs, the mean rate of AHI reduction to <5 is probably close to 50%, and to an AHI <10 is close to 70% [2, 6], but these AHI thresholds for "success" are somewhat arbitrary and do not necessarily correlate with clinical outcome [7]. Higher subjective adherence by a factor of 2 as compared to PAP may account for a similar clinical effectiveness of PAP and MRDs.

OAT can reduce excessive daytime sleepiness and improve alertness, subjectively and objectively [1–37]. Sleepiness has been shown to be objectively reduced as measured by multiple sleep latency testing using an active versus an inactive appliance [8]. Psychomotor vigilance testing performance and driving simulator performance was improved with OAT, even superior to results with CPAP therapy [9, 10]. Not all studies demonstrate reduced sleepiness, especially in mild to moderate OSA [11]. A greater degree of adherence to OAT may account for clinical effectiveness. In comparison, to improve subjective sleepiness using CPAP, >6 h of nightly use may be needed, which represents a level of use not often achieved [12]. Sustained quality of life improvement over 2 years has been demonstrated in MRD-adherent patients with severe OSA. In 103 patients randomized to CPAP or OAT, quality of life as assessed by the FOSQ and SF-36 questionnaires were significantly and similarly improved by both treatments [13].

MRDs may also improve cardiovascular measures. A blood pressure reduction of 2–3 mm Hg has been noted in RCT studies for oral appliances. In another small crossover design study, heart rate variability and nocturnal dipping of blood pressure were shown to be improved using oral appliances for patients with moderate to severe OSA despite only a partial AHI reduction in the majority of patients [14]. Cardiovascular mortality was reduced in an observational study of patients with severe OSA, followed over a median of 79 months, from an odds ratio of 2.1 per 100 person years to odds ratios of 0.61 and 0.56 in those patients treated with MRDs or CPAP, respectively [15].

Mechanism of Action

The predominant mechanism of action is most likely dilation and stabilization of the pharyngeal airway via anatomical connections between the lateral pharyngeal walls and the mandible [16]. In individuals who respond to OAT, there is enlargement of the retropalatal or velopharyngeal airway, mostly in the coronal plane, while the mandible is protruded [17, 18]. This effect of mandibular protrusion can be observed even in the awake patient using fiber-optic endoscopy where cross-sectional area increase has 86% positive and 81% negative predictive values, respectively [19].

Indications

The American Academy of Sleep Medicine (AASM) has recommended that sleep physicians consider prescribing an oral appliance rather than no treatment for adult OSA patients who are either intolerant to or prefer these to CPAP. The recommendation was set at a "standard" level, based on a moderate level of evidence, while noting that PAP "was found to be superior to OAs [oral appliances] in reducing the AHI, arousal index, and oxygen desaturation index and improving oxygen saturation, and therefore, should still generally be the first-line option for treating OSA" [2]. The current practice parameters may reflect both a greater emphasis in patient preference required for a long-term continual therapy, as well as some recognition of recent data on clinical effectiveness. Commercial insurance carriers in the USA do not all follow these guidelines, with some requiring a prior PAP trial of at least 45 days, but most will cover OAT. MRDs may also be used for the treatment of primary snoring [2].

Role of the Otolaryngologist

The otolaryngologist can work collaboratively with a qualified sleep dentist or primarily treat the patient using OAT. The American Academy of Otolaryngology – Head and Neck Surgery supports this with a position statement [38]. MRDs for OSA are not a treatment for the dentition, but rather for a medical disorder. Thus, provided the patient has continual dental care before and during treatment, a trained otolaryngologist may provide OAT. The otolaryngologist may use OAT as the primary therapy for treatment of residual OSA following upper pharyngeal surgery [20–22], or concurrently with PAP if needed. In the USA, the cost to the patient for OAT may be considerably lower when performed by the otolaryngologist, as most participate with the patient's medical insurance and accept insurance payment as full.

Table 1. History and physical factors for oral appliance therapy

History	Physical
Absence of nasal congestion	Absence of obstructive nasal lesions
Stable dentition and ongoing care	Presence of 6–10 healthy teeth per arch
Absence of moderate-severe temporomandibular disorder	Normal jaw mobility
No recent or planned orthodontic therapy	Absence of excess gag response
Bruxism presence or absence	Presence of mandibular protrusion >6 mm
Absence of planned restorations	Absence of periodontal disease, mobile teeth
	Occlusion status: class 2 (favorable)
	Occlusion status: open contact (unfavorable)
	Presence of macroglossia
	Presence of exostoses

Alternatively, an otolaryngologist may work together with a qualified dentist. Fiber-optic endoscopy in the clinic may be used to help with patient selection for success [19, 23]. Nasal obstruction evaluation and treatment can also be beneficial for the OSA patient for use of an appliance that causes at least partial oral obstruction. Nasal obstruction is a predictor of decreased effectiveness of OAT in women [24]. High nasal resistance while supine is also correlated with a nonresponse to OAT [25].

Patient Selection

The first selection is the patient's. The patient must understand that the outcome of OAT is dependent on nightly use of the appliance and must be willing and interested in using an intraoral device. The appliances do not "cure" sleep apnea. The patient's expectations should be assessed and verified as realistic.

Definitive patient history or sleep study criteria to select for a successful OAT outcome are not known. It has been noted by some that women may fare better than men, and that patients who are less obese, have less hypoxemia, lower AHI, and positional-dependent OSA are more likely to succeed [26, 27], but these criteria may simply reflect less severe anatomical disease and are not always predictive or exclusive. Fiber-optic endoscopy of the pharynx with mandibular advancement could be a potential tool to predict treatment success [19, 23], but needs additional validation.

History and Physical Features

Important history and physical factors are listed in Table 1. A nasal congestion history should be elicited and treated prior to OAT as it is associated with treatment failure [24]. Dental care should be confirmed to improve safety and tolerance. The patient should have ongoing dental care with stable dentition and should complete treatment of caries and major restorations prior to engaging in OAT. Oral appliances may reduce dental protection by saliva and could be colonized by bacteria. Periodontal disease should be treated as it may reduce device retention, increase tooth mobility or bite changes, and may worsen with OAT. Bruxism is not a contraindication for OAT but may shorten the appliance lifespan and may influence the choice of appliance. The presence of past temporomandibular dysfunction is not a contraindication. The presence of temporomandibular joint (TMJ) clicking or painless anterior disk displacement with reduction is acceptable.

However, if a patient has active TMJ pain consistent with active inflammation, crepitus consistent with osteoarthritis, or disk displacement without reduction (locking), OAT should be avoided. Patients who have had recent or are planning orthodontic therapy must consider bite changes that may accompany mandibular protrusion.

On examination, there should be 6–10 healthy teeth per arch, in order to provide for adequate retention and allow for the distribution of force along as large an area as possible. Adequate jaw mobility should be ascertained, with jaw opening of 25 mm or more. The status of the dentition, gingiva, and TMJ should be assessed, including occlusion, restorations, wear facets, mobility, and pain. Class 2 occlusion may be advantageous for some as occlusal changes with treatment are more likely to be orthodontically favorable. Patients who underwent 4 premolar extractions for past orthodontic therapy may be more at risk for unfavorable occlusal changes, however. Class 3 occlusion, anterior crossbite, and open contacts are unfavorable as these conditions may worsen with OAT. Exostoses of the jaws may affect the impression process and appliance choice. Tongue size and position should be noted as they may influence appliance choice. Excessive gag may preclude appliance use or warrant appliance modifications to reduce risk. The nasal airway should be assessed for the status of the mucosa and for any obstructive lesions that may mediate obstruction. The patient should have adequate manual dexterity and strength to be able to insert and remove an oral appliance.

The most important patient feature is the ability to protrude the mandible 6 mm or more. With a 6-mm protrusion, oximetry is improved to a greater extent than at lesser protrusion positions [28]. In other studies, AHI outcome was assessed as a function of mandibular protrusion range. A dose-dependent improvement was noted, which was highest at 75% of maximal protrusion as compared with lesser degrees of protrusion [29], or additional protrusion beyond 60% was shown to improve partially treated patients [30]. However, the optimal protrusion position is not known a priori. A temporary, titratable, ready-to-use device called Apnea Guard® can be used prior to custom device fabrication in order to determine the mandibular advancement position. Where available, polysomnographic testing may be performed with mandibular advancement prior to device fabrication [31]. The mandible may be gradually advanced within the patient's range using a commercially available device (MATRx) while using dental impression trays that couple to the MATRx arm for titration. The device has high positive and negative predictive values and can help determine the protrusion range, but requires larger-scale validation [31].

Device Types and Selection

Appliances may be custom made or thermoplastic (boil and bite). Custom appliances are produced from dental impressions at a dental laboratory while thermoplastic appliances are produced immediately in the clinic by heating and molding the device to the patient's dentition. Custom appliances are preferred over thermoplastic ones as per the AASM's consensus recommendations and based on evidence for AHI efficacy [2, 32]. Custom-made appliances may be advantageous as they are more durable, smaller, more comfortable, and may have higher retention and adherence. However, thermoplastic appliances can be easily adjusted in response to changes in the dentition or new restorations, are cheaper, available immediately, and some are quite small and titratable. Titratable appliances are more commonly used as compared with fixed, "monobloc" appliances. Fixed appliances may not be at the optimal position for treatment and may not be easily adjusted if the patient needs further adjustment after the initial efficacy assessment or subsequently.

While MRDs by definition all advance the mandible, there is a plethora of appliance brands

Table 2. Features of oral appliance therapy devices

Chemical composition – polymer type, metallic components
Rigidity – hard, soft, or hybrid rigidity
Thickness and weight
Retention mechanism – friction, clasps
Advancement mechanism – push, pull, bilaterally or midline
Location of hardware – midline, lateral, anterior, posterior
Distribution of forces on dentition and extent of coverage
Tongue space
Advancement range, increments, ease of advancement, vertical adjustment
Allowance for lateral mobility
Allowance for mouth opening
Production process – dental model from impressions versus a 3D-printed model
Laboratory cost

from which to select. Note that some MRDs are only approved for snoring and not OSA, and in the USA the treating doctor should verify the FDA approval status to avoid "off label" use. Appliances may differ in a multitude of factors (Table 2).

Despite the large list, most practitioners select from a handful of devices that accommodate various patient features, including the need for lateral mobility, tongue space, ease of fitting, durability, ease of adjustment, and overall personal experience with a particular brand (Fig. 1). Even with a given appliance, many manufacturers allow for specifying certain modifications on the order sheet to accommodate certain patient or treatment features. For example, one may order hooks for elastic bands to reduce mouth opening on some appliances that normally do not restrict mouth opening.

Side Effects

At the onset of OAT, side effects are common but most resolve within 2 months. They include dental and gingival tenderness, hypersalivation, dry mouth, TMJ discomfort or sounds, myofascial pain, and temporary occlusal changes after a period of sleep. Damage to teeth or restorations may rarely occur. Persistent problems may lead to the discontinuation of treatment, most often related to discomfort or the perceived "cumbersome" nature of ongoing treatment [33]. Occlusal changes are common, but may not be orthodontically unfavorable in class 2 patients [34]. They include a decrease in overjet, overbite, anteroclination of mandibular incisors, and retroclination of the maxillary incisors. Additionally, decreased occlusal contacts and mesial shift of the molar and cuspid occlusion may occur [35]. The degree of change may be related to the extent of mandibular advancement in some studies [35]. Occlusal changes are not a common reason for treatment discontinuation [36]. When exercises are used in the period immediately prior to oral appliance use in sleep, side effects such as TMD pain and temporary bite changes are reduced, sleep and quality of life improve, and OAT adherence may be increased [37].

Clinical Follow-Up and Adherence

The optimal advancement position for a given patient is not known a priori. Therefore, a clinical follow-up and assessment of signs and symptoms is needed. The resolution of disruptive snoring

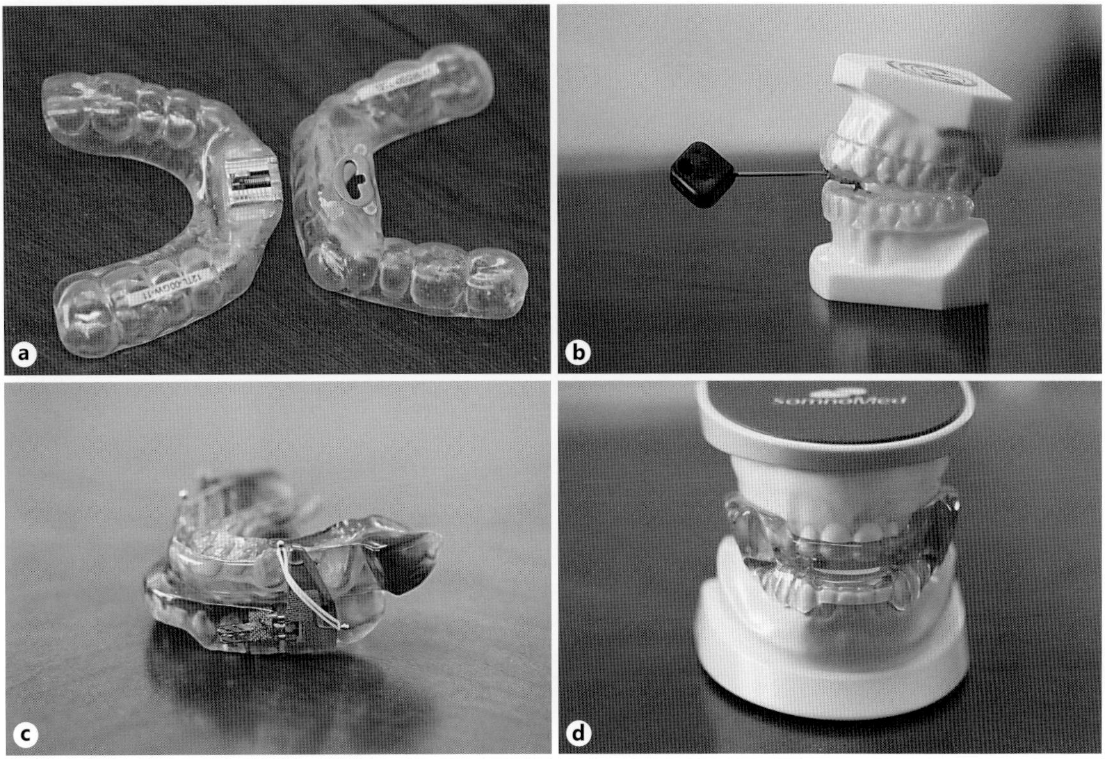

Fig. 1. Examples of oral appliances. **a** TAP-3 appliance with the titration and clasping mechanism shown. **b** TAP-3 titration can be performed with the device in situ. **c** Respire (dorsal fin) device shown with optional hooks for elastics. **d** SomnoDent® device, anterior view with minimal incisor coverage.

and the presence of restorative sleep and daytime function are often considered. Cell phone applications are also available that can acoustically score snoring. For patients with mild OSA, the alleviation of signs and symptoms is the main goal of therapy and AHI reduction may or may not be clinically as relevant. For patients with moderate or severe OSA where the AHI carries significant medical risk, objective sleep testing is needed and can be obtained using portable (limited channel) or in-center polysomnographic testing. If possible, testing should be performed in the treatment position and in another position of greater advancement or of maximally tolerated advancement, which may increase the AHI reduction [30].

Future

Patient selection for a successful outcome remains a challenge. Fiber-optic pharyngoscopy is a promising clinical tool. The extent of mandibular advancement for a given patient is unknown and many begin treatment at 50–60% of maximal protrusion empirically. The use of a remotely controlled mandibular advancement system in sleep centers is promising [31] but may be limited by cost. Titration prior to custom appliance use could be performed at a lower cost using short-term trial appliances (i.e., APneaGuard®) and portable testing.

As 3D printing costs decrease, the availability and affordability of printed OAs may increase.

Such appliances may afford benefits of improved comfort, fit, reduced weight and size, and potentially improved adherence.

Adherence monitors are becoming available (e.g., DentiTrac®) and are important as subjective reporting of adherence is known to overestimate the actual use for CPAP. This may be relevant in particular to patients with severe OSA or for commercial drivers with hypersomnia. Adherence monitors are embedded into the oral appliance and determine usage periods using temperature-based algorithms. However, the reliability of these algorithms may vary. The additional cost of these systems may also present an obstacle toward acceptance.

References

1 Kushida CA, Nichols DA, Holmes TH, Quan SF, Walsh JK, Gottlieb DJ, Simon RD Jr, Guilleminault C, White DP, Goodwin JL, Schweitzer PK, Leary EB, Hyde PR, Hirshkowitz M, Green S, McEvoy LK, Chan C, Gevins A, Kay GG, Bloch DA, Crabtree T, Dement WC: Effects of continuous positive airway pressure on neurocognitive function in obstructive sleep apnea patients: The Apnea Positive Pressure Long-term Efficacy Study (APPLES). Sleep 2012;35: 1593–1602.

2 Ramar K, Dort LC, Katz SG, Lettieri CJ, Harrod CG, Thomas SM, Chervin RD: Clinical practice guideline for the treatment of obstructive sleep apnea and snoring with oral appliance therapy: an update for 2015. J Clin Sleep Med 2015; 11:773–827.

3 Pittman SD, Pillar G, Berry RB, Malhotra A, MacDonald MM, White DP: Follow-up assessment of CPAP efficacy in patients with obstructive sleep apnea using an ambulatory device based on peripheral arterial tonometry. Sleep Breath 2006;10:123–131.

4 Ravesloot MJ, de Vries N, Stuck BA: Treatment adherence should be taken into account when reporting treatment outcomes in obstructive sleep apnea. Laryngoscope 2014;124:344–345.

5 McEvoy RD, Antic NA, Heeley E, Luo Y, Ou Q, Zhang X, Mediano O, Chen R, Drager LF, Liu Z, Chen G, Du B, McArdle N, Mukherjee S, Tripathi M, Billot L, Li Q, Lorenzi-Filho G, Barbe F, Redline S, Wang J, Arima H, Neal B, White DP, Grunstein RR, Zhong N, Anderson CS; SAVE Investigators and Coordinators: CPAP for prevention of cardiovascular events in obstructive sleep apnea. N Engl J Med 2016;375:919–931.

6 Gagnadoux F, Fleury B, Vielle B, Pételle B, Meslier N, N'Guyen XL, Trzepizur W, Racineux JL: Titrated mandibular advancement versus positive airway pressure for sleep apnoea. Eur Respir J 2009; 34:914–920.

7 Weaver EM, Woodson BT, Steward DL: Polysomnography indexes are discordant with quality of life, symptoms, and reaction times in sleep apnea patients. Otolaryngol Head Neck Surg 2005;132: 255–262.

8 Gotsopoulos H, Kelly JJ, Cistulli PA: Oral appliance therapy reduces blood pressure in obstructive sleep apnea: a randomized, controlled trial. Sleep 2004;27: 934–941.

9 Naismith SL, Winter VR, Hickie IB, et al: Effect of oral appliance therapy on neurobehavioral functioning in obstructive sleep apnea: a randomized controlled trial. J Clin Sleep Med 2005;1: 374–380.

10 Phillips CL, Grunstein RR, Darendeliler MA, Mihailidou AS, Srinivasan VK, Yee BJ, Marks GB, Cistulli PA: Health outcomes of continuous positive airway pressure versus oral appliance treatment for obstructive sleep apnea: a randomized controlled trial. Am J Respir Crit Care Med 2013;187:879–887.

11 Marklund M, Carlberg B, Forsgren L, Olsson T, Stenlund H, Franklin KA: Oral appliance therapy in patients with daytime sleepiness and snoring or mild to moderate sleep apnea: a randomized clinical trial. JAMA Intern Med 2015; 175:1278–1285.

12 Weaver TE, Maislin G, Dinges DF, Bloxham T, George CF, Greenberg H, Kader G, Mahowald M, Younger J, Pack AI: Relationship between hours of CPAP use and achieving normal levels of sleepiness and daily functioning. Sleep 2007;30:711–719.

13 Doff MH, Hoekema A, Wijkstra PJ, van der Hoeven JH, Huddleston Slater JJ, de Bont LG, Stegenga B: Oral appliance versus continuous positive airway pressure in obstructive sleep apnea syndrome: a 2-year follow-up. Sleep 2013; 36:1289–1296.

14 Dal-Fabbro C, Garbuio S, D'Almeida V, Cintra FD, Tufik S, Bittencourt L: Mandibular advancement device and CPAP upon cardiovascular parameters in OSA. Sleep Breath 2014;18:749–759.

15 Anandam A, Patil M, Akinnusi M, Jaoude P, El-Solh AA: Cardiovascular mortality in obstructive sleep apnoea treated with continuous positive airway pressure or oral appliance: an observational study. Respirology 2013;18:1184–1190.

16 Brown EC, Cheng S, McKenzie DK, Butler JE, Gandevia SC, Bilston LE: Tongue and lateral upper airway movement with mandibular advancement. Sleep 2013; 36:397–404.

17 Isono S, Tanaka A, Tagaito Y, Sho Y, Nishino T: Pharyngeal patency in response to advancement of the mandible in obese anesthetized persons. Anesthesiology 1997;87:1055–1062.

18 Chan AS, Sutherland K, Schwab RJ, Zeng B, Petocz P, Lee RW, Darendeliler MA, Cistulli PA: The effect of mandibular advancement on upper airway structure in obstructive sleep apnoea. Thorax 2010;65:726–731.

19 Okuno K, Sasao Y, Nohara K, Sakai T, Pliska BT, Lowe AA, Ryan CF, Almeida FR: Endoscopy evaluation to predict oral appliance outcomes in obstructive sleep apnoea. Eur Respir J 2016;47:1410–1419.
20 Millman RP, Rosenberg CL, Carlisle CC, Kramer NR, Kahn DM, Bonitati AE: The efficacy of oral appliances in the treatment of persistent sleep apnea after uvulopalatopharyngoplasty. Chest 1998;113: 992–996.
21 Friedman M, Pulver T, Wilson MN, Golbin D, Leesman C, Lee G, Joseph NJ: Otolaryngology office-based treatment of obstructive sleep apnea-hypopnea syndrome with titratable and nontitratable thermoplastic mandibular advancement devices. Otolaryngol Head Neck Surg 2010;143:78–84.
22 Banhiran W, Kittiphumwong P, Assanasen P, Chongkolwatana C, Metheetrairut C: Adjustable thermoplastic mandibular advancement device for obstructive sleep apnea: outcomes and practicability. Laryngoscope 2014;124: 2427–2432.
23 Sasao Y, Nohara K, Okuno K, Nakamura Y, Sakai T: Videoendoscopic diagnosis for predicting the response to oral appliance therapy in severe obstructive sleep apnea. Sleep Breath 2014;18:809–815.
24 Marklund M, Stenlund H, Franklin KA: Mandibular advancement devices in 630 men and women with obstructive sleep apnea and snoring: tolerability and predictors of treatment success. Chest 2004; 125:1270–1278.
25 Zeng B, Ng AT, Qian J, Petocz P, Darendeliler MA, Cistulli PA: Influence of nasal resistance on oral appliance treatment outcome in obstructive sleep apnea. Sleep 2008;31:543–547.
26 Randerath WJ, Verbraecken J, Andreas S, Bettega G, Boudewyns A, Hamans E, Jalbert F, Paoli JR, Sanner B, Smith I, Stuck BA, Lacassagne L, Marklund M, Maurer JT, Pepin JL, Valipour A, Verse T, Fietze I; European Respiratory Society Task Force on Non-CPAP Therapies in Sleep Apnoea: Non-CPAP therapies in obstructive sleep apnoea. Eur Respir J 2011;37:1000–1028.
27 Chung JW, Enciso R, Levendowski DJ, Morgan TD, Westbrook PR, Clark GT: Treatment outcomes of mandibular advancement devices in positional and nonpositional OSA patients. Oral Surg Oral Med Oral Pathol Oral Radiol Endod 2010;109:724–731.
28 Kato J, Isono S, Tanaka A, Watanabe T, Araki D, Tanzawa H, Nishino T: Dose-dependent effects of mandibular advancement on pharyngeal mechanics and nocturnal oxygenation in patients with sleep-disordered breathing. Chest 2000;117:1065–1072.
29 Aarab G, Lobbezoo F, Hamburger HL, Naeije M: Effects of an oral appliance with different mandibular protrusion positions at a constant vertical dimension on obstructive sleep apnea. Clin Oral Investig 2010;14:339–345.
30 Almeida FR, Parker JA, Hodges JS, Lowe AA, Ferguson KA: Effect of a titration polysomnogram on treatment success with a mandibular repositioning appliance. J Clin Sleep Med 2009;5:198–204.
31 Remmers J, Charkhandeh S, Grosse J, Topor Z, Brant R, Santosham P, Bruehlmann S: Remotely controlled mandibular protrusion during sleep predicts therapeutic success with oral appliances in patients with obstructive sleep apnea. Sleep 2013;36:1517–1525, 1525A.
32 Vanderveken OM, Devolder A, Marklund M, Boudewyns AN, Braem MJ, Okkerse W, Verbraecken JA, Franklin KA, De Backer WA, Van de Heyning PH: Comparison of a custom-made and a thermoplastic oral appliance for the treatment of mild sleep apnea. Am J Respir Crit Care Med 2008;178:197–202.
33 de Almeida FR, Lowe AA, Tsuiki S, Otsuka R, Wong M, Fastlicht S, Ryan F: Long-term compliance and side effects of oral appliances used for the treatment of snoring and obstructive sleep apnea syndrome. J Clin Sleep Med 2005;1:143–152.
34 Almeida FR, Lowe AA, Otsuka R, Fastlicht S, Farbood M, Tsuiki S: Long-term sequellae of oral appliance therapy in obstructive sleep apnea patients: part 2. Study-model analysis. Am J Orthod Dentofacial Orthop 2006;129:205–213.
35 Doff MH, Finnema KJ, Hoekema A, Wijkstra PJ, de Bont LG, Stegenga B: Long-term oral appliance therapy in obstructive sleep apnea syndrome: a controlled study on dental side effects. Clin Oral Investig 2013;17:475–482.
36 Fritsch KM, Iseli A, Russi EW, Bloch KE: Side effects of mandibular advancement devices for sleep apnea treatment. Am J Respir Crit Care Med 2001;164:813–818.
37 Cunali PA, Almeida FR, Santos CD, Valdrichi NY, Nascimento LS, Dal-Fabbro C, Tufik S, Bittencourt LR: Mandibular exercises improve mandibular advancement device therapy for obstructive sleep apnea. Sleep Breath 2011;15:717–727.
38 American Academy of Otolaryngology – Head and Neck Surgery: Position statement: use of oral appliances for the treatment of obstructive sleep apnea (OSA). 2014. http://www.entnet.org/?q=node/216.

Ofer Jacobowitz, MD, PhD, FAASM
ENT and Allergy Associates
261 5th Avenue, Suite 901
New York, NY 10016 (USA)
E-Mail ojacobowitz@entandallergy.com

Lin H-C (ed): Sleep-Related Breathing Disorders. Adv Otorhinolaryngol. Basel, Karger, 2017, vol 80, pp 66–73
DOI: 10.1159/000470868

Updated Nasal Surgery for Obstructive Sleep Apnea

David M. Johnson • Ryan J. Soose

Department of Otolaryngology, University of Pittsburgh Medical Center, Pittsburgh, PA, USA

Abstract

Numerous studies report that difficulty breathing through the nose is associated with reduced sleep quality, increased daytime dysfunction, and increased risk of snoring and obstructive sleep apnea (OSA). Additionally, chronic nasal obstruction often complicates and limits successful medical device therapy for OSA, particularly with continuous positive airway pressure (CPAP). When medical evaluation and management of the nose is inadequate, surgical therapy to lower nasal resistance has been shown to substantially improve sleep and breathing outcome measures. Although nasal surgery in isolation does not have a consistent effect on the apnea-hypopnea index in OSA patients, it does have robust evidence on improving snoring, subjective sleep quality, daytime sleepiness, sleep-related quality of life measures, and other important OSA outcome measures. Furthermore, in OSA patients with sinonasal disease who are struggling with CPAP and other medical devices, sinonasal surgery also has the potential to lower pressure requirements and improve adherence rates. Nasal surgery plays a key adjunctive role in the management of an OSA patient population.

Epidemiology

The nasal airway and sleep-disordered breathing (SDB) are closely linked. Experimentally induced nasal obstruction in healthy subjects led to changes in sleep architecture with increased arousals and apneas [1–3]. In large cohort studies, nasal obstruction, as determined by rhinometry or patient self-report, has been shown to be an independent risk factor for snoring and for obstructive sleep apnea (OSA); indeed, among subjects reporting severe chronic nasal congestion, the odds ratio of habitual snoring was 3.6 at baseline and 4.9 at the 5-year follow-up [4–7]. Treatment of nasal inflammatory conditions such as allergic rhinitis is known to improve subjective sleep and breathing outcomes, although the evidence for improvement in polysomnography findings is not as robust [8].

Pathophysiology

The effects of nasal airflow on sleep physiology are complex. In awake subjects with normal nasal resistance, an average of 8% of ventilatory volume is taken by mouth; this decreases to just 4% during all stages of sleep [9]. The nose accounts for at least half of total airway resistance, and by the Starling resistor model, if this upstream proportion is increased, the likelihood of downstream pharyngeal collapse is higher. Although among laypersons the nose is commonly thought to be the source of nocturnal stertor, in the majority of snorers it is the uvula, soft palate, and pharyngeal soft tissues that are responsible. Although the source of snoring is most commonly the soft palate and other pharyngeal soft tissues, increased nasal resistance can exacerbate snoring inasmuch as the faster airflow velocity generated by a narrowed nasal airway may create a Bernoulli effect that results in, or worsens, soft tissue fluttering. Increased resistance in the nose can also lead to increased respiratory effort, generating negative intrathoracic pressures that exacerbate collapsibility throughout the upper airway.

Another misconception is that open-mouth breathing effectively improves nighttime respiration by bypassing nasal obstruction. In fact, an open-mouth posture retrudes the mandible, allowing the base of the tongue to narrow the oropharyngeal airway, resulting in greater airway resistance overall [10]. Furthermore, nasal afferents are associated with reflexes (perhaps including the putative "nasopulmonary reflex") that triggers bronchodilation [11]. Nitric oxide produced by the sinonasal mucosa in response to nasal airflow stimulates pulmonary vasodilation and improves ventilation-perfusion ratios [12]. Nasal blockage likely attenuates the airway-related reflex pathways and reduces the transport of nitric oxide to the lungs, worsening V-Q mismatch. Temporary chemical denervation of the nasal and nasopharyngeal mucosa in normal subjects led to increased obstructive and central SDB events [13]. Thus, blockage of normal nasal airflow has a cascading effect on SDB.

Increased nasal resistance can stem from inflammatory mucosal conditions or anatomic abnormalities (Table 1). A thorough history and examination, along with imaging or endoscopy as indicated, can often differentiate between these etiologies, as the former are typically managed medically whereas the latter may benefit from surgical intervention. Symptomatic nasal obstruction is often multifactorial and may require combination medical and surgical therapy for the optimum long-term management of the nasal airway. The approach in these cases should be tailored to the individual patient, guided by the symptomatic response to treatment [14]. Patients with moderate to severe OSA are more likely than controls to demonstrate sinonasal disorders, which in turn may limit successful OSA management with positive pressure therapy and other OSA medical devices [15]. In patients with both OSA and sinonasal disease, surgical treatment of the nose has the potential not only to improve sleep and breathing outcome measures, but also to improve success rates with OSA medical therapy.

Evaluation of the Nasal Airway

A complete sinonasal history and thorough upper airway exam are warranted in any SDB patient for whom surgical intervention is considered. Pertinent history includes facial trauma, prior surgery (particularly sinonasal surgery including cosmetic rhinoplasty, tonsillectomy and adenoidectomy, and any procedures on the tongue, pharynx, larynx, or trachea), allergic rhinitis, cardinal chronic rhinosinusitis symptoms, nasal polyps, dental complaints, asthma and chronic obstructive pulmonary disease, and aspirin sensitivity. A validated instrument, such as the Sinonasal Questionnaire or Nasal Obstruction Symptom Evaluation, may be helpful in identifying patients with sino-

Table 1. Selected causes of nasal obstruction [adapted from 8]

Medical	Structural
Allergic rhinitis	Deviated nasal septum
Vasomotor rhinitis	Inferior turbinate hypertrophy
Atrophic rhinitis	Acquired nasal deformity
Rhinitis medicamentosa	Nasal valve collapse
Nonallergic rhinitis with eosinophilia syndrome	Concha bullosa
Medications (e.g., aspirin, oral contraceptives, antihypertensives)	Nasal polyps
Infection (e.g., common cold, acute bacterial sinusitis)	Adenoid hypertrophy
Hormonal (e.g., pregnancy, hypothyroidism)	Tonsillar hypertrophy
Autoimmune (e.g., granulomatosis with polyangiitis, lupus, rheumatoid arthritis)	Narrow pyriform aperture
Occupational (e.g., chemicals, paint fumes, chalk dust, sawdust, perfumes)	Inverting papilloma
Smoking	Neoplasm
	Foreign body
	Synechiae
	Meningocele
	Granulomatous disease

nasal concerns and to longitudinally assess the response to therapy [16–17] (Fig. 1, 2).

Examination begins with an inspection of the external nose. Symmetry and the width of the dorsum and alae are assessed at rest, during quiet breathing, and with deep inspiration. Nasal valve insufficiency is tested with the Cottle maneuver: the examiner uses a finger to stretch the skin next to each ala laterally; the patient respires and notes whether nasal blockage is subjectively improved. The specificity of the Cottle maneuver can be improved by using a cerumen curette to retract the upper and lower lateral cartilages independently.

Anterior rhinoscopy with a nasal (Vienna) speculum and bright headlight provides adequate visualization for many patients. The examiner braces the speculum hand on the patient's cheek or nose and opens the tines vertically so they do not contact the exquisitely sensitive septal mucosa. The vestibule and nasal cavity are systematically inspected, with attention to the configuration of the septum (intact/perforated, location and direction of any deviation), size of the inferior and middle turbinates, color and condition of the mucosa, and presence of any polyp, crusting, or discharge.

Multilevel obstruction is common in SDB patients, and a complete head and neck exam is mandatory prior to surgical decision making. Particular attention must be paid to the tonsils (palatine and lingual) and adenoid pad, as the presence of excessive lymphoid tissue in the Waldeyer ring causing posterior obstruction renders moot any intranasal work. The adenoids are examined with a laryngeal mirror and headlight, or with a nasal endoscope. The endoscope and selective imaging are indispensable in cases where significant sinonasal disease (chronic rhinosinusitis, polyposis, neoplasm) is suspected.

Evidence for Surgical Therapy

Altering nasal resistance surgically has been shown to improve nasal resistance and subjective sleep-related symptoms. In OSA patients with a deviated septum and symptoms of nasal obstruction, Li et al. [18] compared septoplasty with inferior turbinate reduction to medical therapy

Over the last 3 months how often, on average, did you have the following symptoms?

	Never	1–4 times per month	2–6 times per week	Daily
Runny nose	☐	☐	☐	☐
Postnasal drip	☐	☐	☐	☐
Need to blow your nose	☐	☐	☐	☐
Facial pain/pressure	☐	☐	☐	☐
Nasal obstruction	☐	☐	☐	☐

Scoring: never (0), 1–4 times per month (1), 2–6 times per week (2), and daily (3).

Score reported as average of 5 items: range of possible scores 0–3.

Fig. 1. Sinonasal Questionnaire (SNQ). Originally designed as a tool to screen for chronic sinonasal disease in an asthma population, the SNQ is a simple instrument that may facilitate initial screening and longitudinal follow-up of sinonasal problems in sleep-disordered breathing patients [17; with permission from SAGE Publications, Inc.].

(steroid or decongestant spray, saline lavage, or oral antihistamine) in 66 nonpathologically obese patients (BMI <33) with an apnea-hypopnea index (AHI) >5 in a prospective, controlled but nonrandomized trial. Statistically significant differences were demonstrated in the Snore Outcome Survey (SOS), Spouse/Bed Partner Survey (SBPS), Epworth Sleepiness Scale (ESS), and rhinomanometric measurement of nasal resistance. No significant change in AHI was obtained in the main analysis; however, when stratified into various subgroups, a significant effect of nasal surgery on AHI was found when patients were divided by Friedman Tongue Position (FTP) into "low" (FTP 1 or 2) and "high" (FTP 3 or 4) groups. These findings are consistent with previously published reports that correlate AHI reduction after nasal surgery with favorable pharyngeal anatomy characteristics (e.g., small tonsils, low modified Mallampati or FTP score, and a normal mandibular plane to hyoid distance on cephalometry). In contrast, increased airway length (i.e., low hyoid position), large tonsils, and high modified Mallampati or FTP score has been associated with a lack of AHI improvement after nasal surgery in isolation.

The same group then conducted a meta-analysis that confirmed the robustness of these findings [19]. Based on 9 prospective, noncontrolled trials, the weighted mean AHI prior to nasal surgery was 35.2 ± 22.6, and postoperatively was 33.5 ± 23.8, yielding $p = 0.69$. However, subjective measures of OSA symptoms decreased significantly: ESS from 10.6 ± 3.9 to 7.1 ± 3.7 ($p < 0.001$) in 9 noncontrolled studies, and in the single case-control study analyzed there was a similar de-

Nasal Obstruction Symptom Evaluation (NOSE) Instrument

→ **To the Patient:** Please help us to better understand the impact of nasal obstruction on your quality of life by **completing the following survey**. Thank You!

Over the past 1 month, how much of a problem were the following conditions for you?

Please circle the most correct response

	Not a problem	*very mild problem*	*moderate problem*	*fairly bad problem*	*severe problem*
1. Nasal congestion or stuffiness	0	1	2	3	4
2. Nasal blockage or obstruction	0	1	2	3	4
3. Trouble breathing through my nose	0	1	2	3	4
4. Trouble sleeping	0	1	2	3	4
5. Unable to get enough air through my nose during exercise or exertion	0	1	2	3	4

Otolaryngology–Head and Neck Surgery

Fig. 2. Nasal Obstruction Symptom Evaluation (NOSE) scale. The NOSE scale is another validated and simple tool that may be used to more specifically evaluate and monitor symptomatic nasal obstruction in sleep-disordered breathing patients [17; with permission from SAGE Publications, Inc.].

crease in the surgery group but not in the control group, suggesting that the placebo effect does not fully account for the change. Measures of snoring (SBPS, SOS, and visual analog scale for snoring) were also reduced postoperatively, but with such variations in improvement that the authors posited multiple confounding factors (age, gender, obesity, upper airway anatomy including tonsil size, etc.).

Since the publication of the meta-analysis, a number of other studies have expanded the literature. An intriguing investigation from Japan studied compound nasal surgery (septoplasty, turbinate reduction, and submucosal resection of the posterior nasal nerve) in 53 patients with high nasal resistance and moderate or severe OSA, finding that 57 and 75%, respectively, showed a postoperative improvement in at least 1 polysomnography parameter [20]. Quality of life measures were also significantly improved. Posterior nasal neurectomy hypothetically reduces cholinergic input to the turbinates, inhibiting mucosal swelling; perhaps it is this additional contribution to lowered nasal airway resistance that leads to positive consequences on sleep apnea. In the plastic surgery literature, a prospective trial of functional septorhinoplasty without turbinate reduction was undertaken in the setting of OSA [21]. Overall, AHI scores improved from 24.7 to 16 ($p = 0.013$); participants whose BMI was below 30 improved from 22.5 to 9.6 ($p < 0.01$). The authors hypothesized that other studies of nasal airway surgery failed to improve AHI due to an inadequate extent of nasal valve surgery. Additionally, patients with mild to moderate positional OSA may respond better to nasal surgery than those with nonpositional OSA [22].

Despite the above evidence, the importance of lowering nasal resistance in OSA patients has been understated in the past because of the variable effect of nasal surgery on the AHI. Although the AHI remains an important surrogate marker for cardiovascular risk and an objective outcome measure, sleep physicians must remember that the AHI is only 1 metric of a complex and heterogeneous condition. For example, Nakata et al. [23] reported that nasal surgery in patients with OSA and symptomatic nasal obstruction improved nasal resistance, sleep architecture, and daytime sleepiness, including a reduction in the apnea-hypopnea duration. In a prospective longitudinal cohort study of OSA patients with nasal obstruction, nasal surgery has also been shown to improve disease-specific and general health (SF-36 health survey) quality of life measures [24]. Finally, even in patients without clinically significant OSA, nasal surgery to treat structural nasal obstruction was associated with improvement in the Pittsburgh Sleep Quality Index (PSQI), further emphasizing the importance of a good nasal airway on sleep parameters well beyond the AHI assessment alone [25].

Nasal airway surgery also has an important adjunctive role in the ability of patients to adhere to and accommodate OSA medical device therapy, including continuous positive airway pressure (CPAP). Increased nasal resistance has been associated with initial nonacceptance of CPAP after titration [26]. Nasal resistance was also 1 of only 2 independent predictors of poorer outcomes with oral appliance therapy for OSA [27]. Up to half of CPAP users report symptomatic nasal obstruction and, although a full facemask may theoretically bypass nasal resistance, full face or oronasal interfaces have been associated with higher CPAP pressure requirements and lower adherence rates [28]. The treatment of lower nasal resistance and facilitation of successful nasal interface application may provide an opportunity to improve adherence and CPAP outcomes. A recent meta-analysis examined 18 studies that reported CPAP data after nasal surgery alone [29]. Seven reported pre- and postoperative mean airway pressures and standard deviations; a statistically significant effect from 11.6 ± 2.2 to 9.5 ± 2.0 cm H_2O was demonstrated in a pooled cohort of 82 patients. CPAP device downloads showed an increase from 3.0 ± 3.1 to 5.5 ± 2.0 h of CPAP use per night, although study heterogeneity may limit the applicability of these data. In 11 studies that contained patient-reported CPAP usage estimates, subgroup analysis on 64 patients who did not tolerate CPAP preoperatively yielded 89% acceptance of CPAP postoperatively.

Finally, structural causes of increased nasal resistance are not limited to the nasal cavity. Studies have shown improvement in nasal resistance of pediatric and adult OSA patients after adenotonsillectomy and tonsillectomy alone [30]. A detailed review of these data are beyond the scope of this discussion; however, in general if a patient is to undergo sinonasal surgery for SDB, hypertrophied tonsils and adenoids should also be treated. Pediatric SDB patients with adenotonsillar hypertrophy and turbinate hypertrophy benefit from inferior turbinate reduction concomitantly with adenotonsillectomy, demonstrating subjective and polysomnographic evidence of greater improvement than control patients with adenotonsillar hypertrophy and turbinate hypertrophy who underwent adenotonsillectomy alone [31].

Conclusion

The management of nasal abnormalities is a key adjunctive treatment in patients with OSA. Thorough evaluation of the nasal airway and thoughtful application of selected surgical interventions can significantly improve patient-reported outcomes and success with OSA medical therapy, including CPAP.

References

1 Olsen KD, Kern EB, Westbrook PR: Sleep and breathing disturbance secondary to nasal obstruction. Otolaryngol Head Neck Surg 1981;89:804–810.
2 Lavie P, Fischel N, Zomer J, Eliaschar I: The effects of partial and complete mechanical occlusion of the nasal passages on sleep structure and breathing in sleep. Acta Otolaryngol 1983;95:161–166.
3 Zwillich CW, Pickett C, Hanson FN, Weil JV: Disturbed sleep and prolonged apnea during nasal obstruction in normal men. Am Rev Respir Dis 1981;124:158–160.
4 Lofaso F, Coste A, d'Ortho MP, Zerah-Lancner F, Delclaux C, Goldenberg F, Harf A: Nasal obstruction as a risk factor for sleep apnoea syndrome. Eur Respir J 2000;16:639–643.
5 Young T, Finn L, Palta M: Chronic nasal congestion at night is a risk factor for snoring in a population-based cohort study. Arch Intern Med 2001;161:1514–1519.
6 Young T, Finn L, Kim H: Nasal obstruction as a risk factor for sleep-disordered breathing. The University of Wisconsin Sleep and Respiratory Research Group. J Allergy Clin Immunol 1997;99:S757–S762.
7 Stradling JR, Crosby JH: Predictors and prevalence of obstructive sleep apnoea and snoring in 1,001 middle aged men. Thorax 1991;46:85–90.
8 Kent DT, Soose RJ: Environmental factors that can affect sleep and breathing: allergies. Clin Chest Med 2014;35:589–601.
9 Fitzpatrick M, Driver H, Chatha N, et al: Partitioning of inhaled ventilation between the nasal and oral routes during sleep in normal subjects. J Appl Physiol 2003;94:883–890.
10 Fitzpatrick MF, McLean H, Urton AM, et al: Effect of nasal or oral breathing route on upper airway resistance during sleep. Eur Respir J 2003;22:827–832.
11 Sato T: Effect of nasal mucosa irritation on airway resistance. Auris Nasus Larynx 1980;7:39–50.
12 Blitzer ML, Loh E, Roddy MA, et al: Endothelium-derived nitric oxide regulates systemic and pulmonary vascular resistance during acute hypoxia in humans. J Am Coll Cardiol 1996;28:591–596.
13 White DP, Cadieux RJ, Lombard RM, et al: The effects of nasal anesthesia on breathing during sleep. Am Rev Respir Dis 1985;132:972.
14 Prokopakis E, Vlastos I, Pant H, Ferguson BJ: Chronic rhinosinusitis: observation or treatment. Curr Opin Allergy Clin Immunol 2013;13:31–36.
15 de Aguiar Vidigal T, Martinho Haddad FL, Gregorio LC, et al: Subjective, anatomical, and functional nasal evaluation of patients with obstructive sleep apnea syndrome. Sleep Breath 2013;17:427–433.
16 Dixon AE, Sugar EA, Zinreich SJ, et al: Criteria to screen for chronic sinonasal disease. Chest 2009;136:1324–1332.
17 Stewart MG, Witsell DL, Smith TL, et al: Development and validation of the Nasal Obstruction Symptom Evaluation (NOSE) scale. Otolaryngol Head Neck Surg 2004;130:157–162.
18 Li HY, Lee LA, Wang PC, Fang TJ, Chen NH: Can nasal surgery improve obstructive sleep apnea: subjective or objective? Am J Rhinol Allergy 2009;23:e51–e55.
19 Li H, Wang P, Chen Y, et al: Critical appraisal and meta-analysis of nasal surgery for obstructive sleep apnea. Am J Rhinol Allergy 2011;25:45–49.
20 Hisamatsu K, Kudo I, Makiyama K: The effect of compound nasal surgery on obstructive sleep apnea syndrome. Am J Rhinol Allergy 2015;29:e192–e196.
21 Shuaib SW, Undavia S, Lin J, Johnson CM Jr, Stupak HD: Can functional septorhinoplasty independently treat obstructive sleep apnea? Plast Reconstr Surg 2015;135:1554–1565.
22 Hu B, Han D, Li Y, Ye J, Zang H, Wang T: Polysomnographic effect of nasal surgery on positional and non-positional obstructive sleep apnea/hypopnea patients. Acta Otolaryngol 2013;133:858–865.
23 Nakata S, Noda A, Yasuma F, et al: Effects of nasal surgery on sleep quality in obstructive sleep apnea syndrome with nasal obstruction. Am J Rhinol 2008;22:59–63.
24 Li HY, et al: Improvement in quality of life after nasal surgery alone for patients with obstructive sleep apnea and nasal obstruction. Arch Otolaryngol Head Neck Surg 2008;134:429–433.
25 Stapleton AL, Chang Y, Soose RJ, Gillman GS: The impact of nasal surgery on sleep quality: a prospective outcomes study. Otolaryngol Head Neck Surg 2014;151:868–873.

26 Suguira T, Noda A, Nakata S, et al: Influence of nasal resistance on initial acceptance of CPAP in treatment of OSAS. Respiration 2007;74:56–60.

27 Zeng B, Ng AT, Qian J, et al: Influence of nasal resistance on oral appliance treatment outcome in obstructive sleep apnea. Sleep 2008;31:543–547.

28 Andrade RG, Piccin VS, Nascimento JA, et al: Impact of the type of mask on the effectiveness of and adherence to CPAP treatment for obstructive sleep apnea. J Bras Pneumol 2014;40:658–668.

29 Camacho M, Riaz M, Capasso R, Ruoff CM, Guilleminault C, Kushida CA, Certal V: The effect of nasal surgery on continuous positive airway pressure device use and therapeutic treatment pressures: a systematic review and meta-analysis. Sleep 2015;38:279–286.

30 Nakata S, Miyazaki S, Ohki M, et al: Reduced nasal resistance after simple tonsillectomy in patients with OSA. Am J Rhinol 2007;21:192–195.

31 Sullivan S, Kasey Li, Guilleminault C: Nasal obstruction in children with sleep-disordered breathing. Ann Acad Med Singapore 2008;37:645–648.

Ryan J. Soose, MD
UPMC Mercy Building D, Suite 2100
1400 Locust Street
Pittsburgh, PA 15219 (USA)
E-Mail soosrj@upmc.edu

Lin H-C (ed): Sleep-Related Breathing Disorders. Adv Otorhinolaryngol. Basel, Karger, 2017, vol 80, pp 74–80
DOI: 10.1159/000470869

Updated Palate Surgery for Obstructive Sleep Apnea

Hsueh-Yu Li

Department of Otolaryngology, Chang Gung Memorial Hospital, School of Medicine, Chang Gung University, Taoyuan, Taiwan, ROC

Abstract

Initially described in 1981, uvulopalatopharyngoplasty was the first surgical procedure specially designed at the palatal level for snoring and obstructive sleep apnea (OSA). To date, palatal surgery remains the most commonly used surgery for OSA. The advancement of this surgery over the past 30 years has been a process of evolution in concepts, examination, and technique to increase safety and effectiveness in the treatment of OSA. Concept changes have emerged in the disease etiology, purpose of surgery, treatment priorities, staging of operations, integration therapy, and surgical endpoints. Drug-induced sleep examination has become the mainstream for forming a surgical plan. The surgical technique has tended toward the functional expansion and stabilization of airway tissue instead of excision. Here, drug-induced sleep computed tomography is introduced. In addition, palatal surgery is further divided into palatoplasty and pharyngoplasty for individual clinical application.

Continuous positive airway pressure (CPAP), oral appliance, and surgery are mainstream approaches in the treatment of obstructive sleep apnea (OSA). Among them, surgery is the only treatment modality to improve OSA without the use of a device. Uvulopalatopharyngoplasty (UPPP), as described by Fujita et al. [1] in 1981, was the first surgical procedure at the palatal level specifically designed to treat snoring and obstructive sleep apnea (OSA). Originally, UPPP included excision of the tonsils, trimming and relocating the tonsillar pillars, uvulectomy, and mucosal closure of the soft palate. Although UPPP improved the symptoms of OSA, it was criticized for its low success rate and severe postoperative pain and related complications [2]. Over the past 30 years, UPPP or palatal surgery has remained the most commonly used operation for snoring and OSA. However, there have been many changes in palatal surgery for OSA in terms of the concept, examination to determine the obstruction site, and operation technique.

Concept Changes

Disease Etiology

The etiology of OSA could be both anatomical obstruction and neurological collapse, with different proportions in individual patients. Anatomical obstruction, such as tonsillar hypertrophy, is an obvious phenotype characteristic; neurological collapse, such as weakness of muscle tone, may be attributable to the genotype. The result of palatal surgery for OSA is sometimes unpredictable and partially based on the proportion between the 2 types; it usually has better outcomes in treating anatomical obstruction than neurological collapse. Our research has revealed that anatomy staging is more reliable than severity staging in predicting outcomes from palatal surgery for OSA [3].

Surgical Purpose

Evidence shows that palatal surgery improves OSA-related symptoms (such as snoring, daytime sleepiness, observed sleep apnea, and quality of life), decreases coronary heart/cardiovascular disease and mortality, but rarely normalizes the apnea/hypopnea index in severe OSA patients [4–6]. Furthermore, the incidence and severity of OSA is related to age. OSA is likely to relapse with an increase in age. Therefore, it is not realistic to maintain the normalization of the apnea-hypopnea index (AHI) over a patient's entire lifetime. The purpose of palatal surgery thereby changes from cure to an improvement of quality of life, decrease of disease severity, and prevention of major complications in middle age.

Treatment Priority

CPAP is usually suggested as the first line of treatment for OSA [7]. It is most suitable in OSA patients who are obese, elderly, have severe severity, daytime sleepiness, or have had a previous heart attack or cardiovascular event. By contrast, CPAP therapy may have poor compliance/adherence and low acceptance in young, nonobese, nonsleepy, and mild OSA patients. Palatal surgery takes the precedence over CPAP in the aforementioned patients, especially those with large tonsils, to improve their snoring, quality of life, and disease severity.

Classification of Sleep Surgery

Soft tissue surgery, bone surgery, and bariatric surgery comprise sleep surgery. Soft tissue surgery is considered "local" surgery in tissue volume reduction and enlargement of the airway lumen. Palatal surgery plays a key role in soft tissue surgery and has the advantage of directly treating both partial obstruction (snoring) and total obstruction (apnea) of the upper airway. Bone surgery, such as maxillomandibular advancement, is regarded as "regional" surgery and the last resort of sleep surgery to tense the lateral pharyngeal wall and enlarge both the retropalatal and retrolingual space in severe OSA patients [8]. Bariatric surgery is deemed "systemic" surgery for the general reduction of adipose tissue in pathologically obese patients [9].

Treatment Modality

Since the etiology of OSA is multifactorial, single-level surgery (palatal surgery) usually causes suboptimal results. An integration of palatal surgery in multilevel surgery with positional therapy and oropharyngeal muscle training improves treatment outcomes in both the short- and long-term follow-up [10–14].

Endpoints

Research has revealed that posttreatment AHI is not consistent in improving clinical symptoms (representing a major concern for patients) and patient satisfaction. Thus, the endpoint of palatal surgery is not fully concerned with the normalization of AHI and shifts from a single-value AHI to a comprehensive and holistic interpretation. The following combination is suggested as the endpoint of palatal surgery for OSA: (1) a significant improvement or vanishing of major complaints,

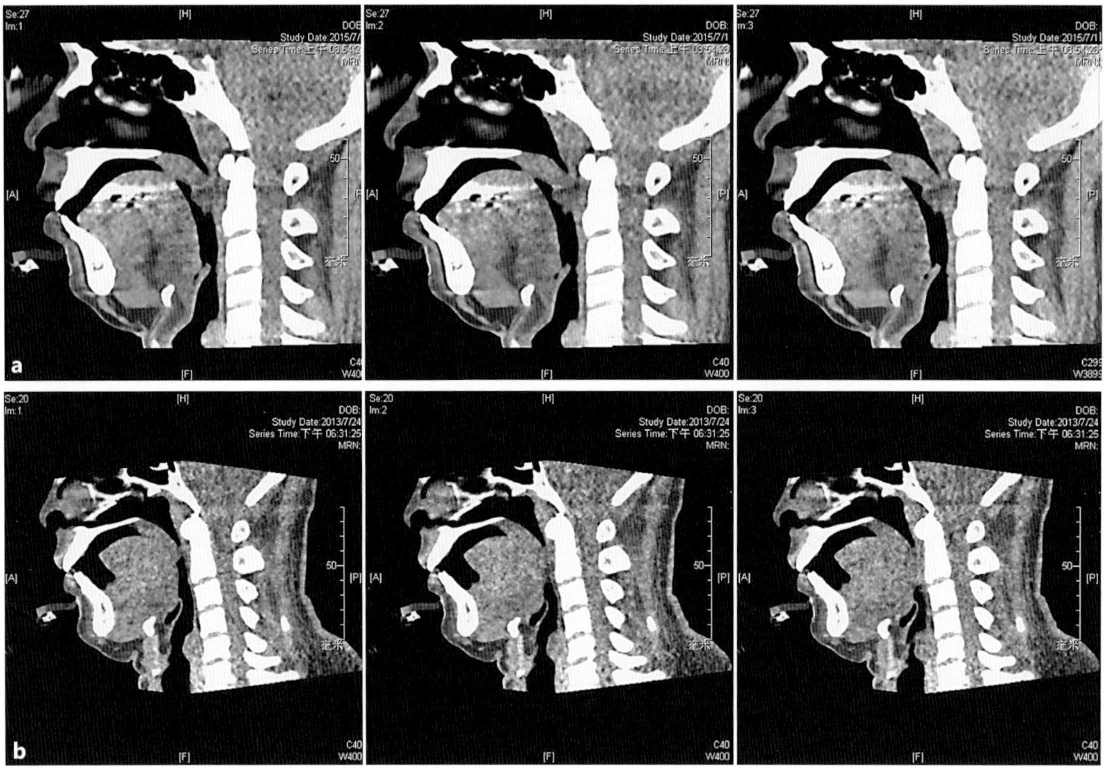

Fig. 1. Drug-induced sleep computed tomography. **a** Pure palatal obstruction. **b** Tongue collapse leading to palatal obstruction.

(2) satisfaction of the patient, (3) AHI <15, and (4) a low risk in biomarkers of coronary artery/cardiovascular disease [10, 15].

Changes in Examination

Conventional evaluation of airway obstruction for palatal surgery in treating OSA includes the Friedman stage, nasopharyngoscopy, and cephalometry, which are generally implemented in wakefulness. However, evidence shows that obstruction sites detected during wakefulness are not identical to those detected during sleep. Therefore, drug-induced sleep endoscopy (DISE) has become a commonly used preoperative examination for surgical OSA patients [16]. The advantage of DISE is the provision of dynamic upper airway collapse during pharmacologically induced sleep. However, its drawbacks from intraluminal examination may include only one-level obstruction observed during one site evaluation, misdiagnosis of anterior-posterior palatal obstruction from tongue compression, and difficulty in assessing the precise obstruction area of the tongue.

By the reinforcement in speed of computed tomography scanning, fast dynamic movement of the upper airway during sedation sleep becomes another option for assessing surgical OSA patients. Drug-induced sleep imaging can depict dynamic airway movement, especially in the sagittal view. This demonstrates the interaction of the palate/tongue obstruction and the anterior-

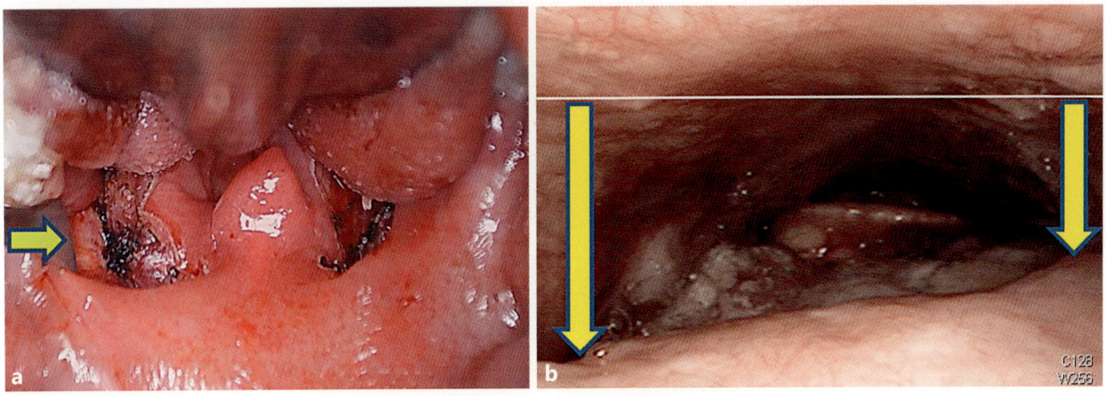

Fig. 2. a Pterygomandibular raphe (arrow). **b** Difference in the velopharyngeal space after the operation. The long arrow indicates the fixation of the palatopharyngeal muscle to the pterygomandibular raphe; the short arrow shows the traditional uvulopalatopharyngoplasty.

posterior obstruction area in the palate and tongue (Fig. 1), which could be helpful in formulating a surgical plan and improving treatment outcomes [17].

Changes in Surgical Technique

The history of palatal surgery for OSA reveals an evolutionary process. Many modifications in surgical technique have been implemented to minimize the side effects on pharyngeal function and maximize the surgical effect to sleep apnea [18–28].

Palatal Surgery Selection Based on the Pattern of Collapse

Supratonsillar palatoplasty (part of relocation pharyngoplasty) is suggested for treating anterior-posterior collapse of the soft palate. The operation is initiated with a bilateral tonsillectomy. Two elliptical cuts over the supratonsillar area are made. Tissue in this semilunar area, including mucosa and submucosal adipose tissue, is dissected from the underlying muscles and removed. The posterior pillar flap is everted and sewn onto the residual anterior pillar in the style of a mattress. The mattress suture starts from a low position and is located inside to facilitate pushing of the mucosal flap in an anterolateral direction. The distal (nonmuscle) part of the uvula is excised. This technique is a modification of extended uvulopalatoplasty [20].

Relocation pharyngoplasty is implemented to improve collapse from the lateral pharyngeal wall. The procedure is performed following a supratonsilloplasty. The superior pharyngeal constrictor muscle muscle is grasped and sewn laterally to the ipsilateral palatoglossus muscle in a cephalic-to-caudal direction with a 2-0 vicryl mattress-style suture. The suture usually requires 3–4 separated stitches to splint the superior pharyngeal constrictor muscle in 1 tonsillar fossa. The posterior pillar flap is then erected from a horizontal to vertical plane and sewn onto the residual palatoglossus muscle. The same procedure is repeated on the opposite side [24].

Functional Expansion with Anchor in the Pterygomandibular Raphe

The main cause of failure in UPPP can be attributed to untreated tongue/epiglottis obstruction

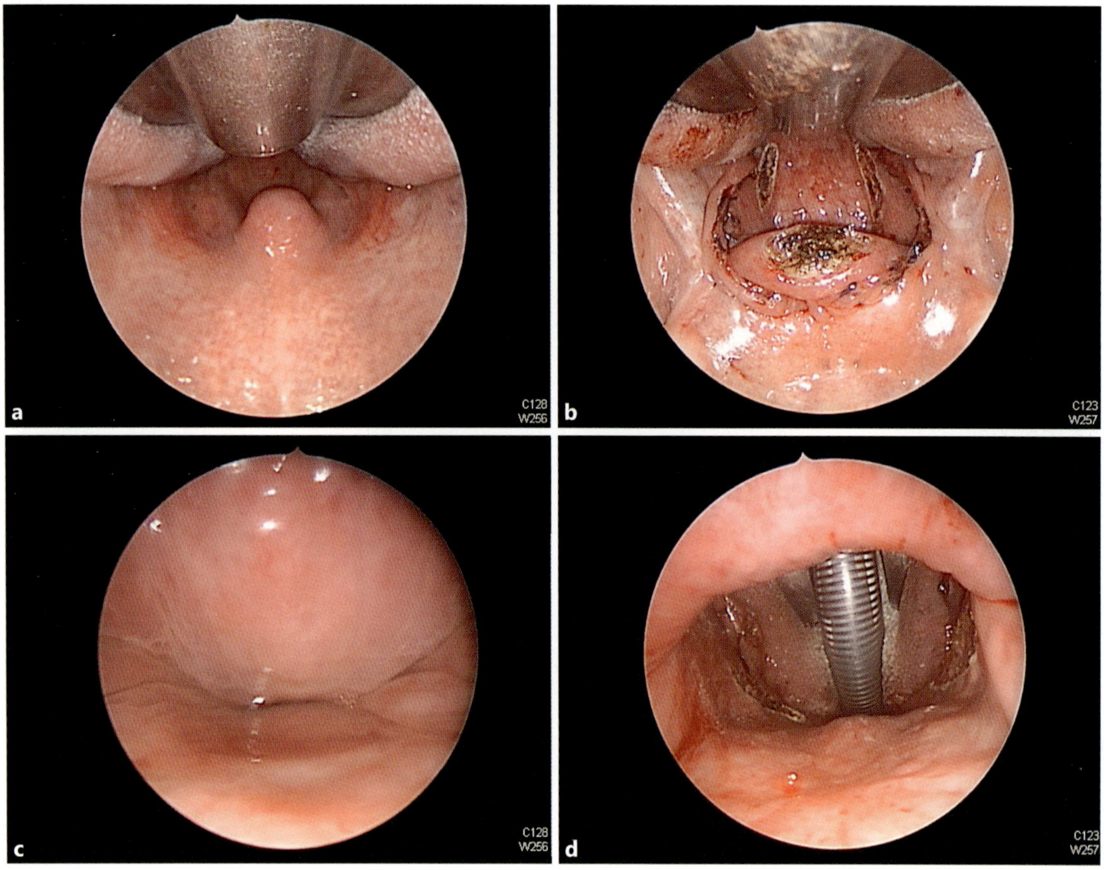

Fig. 3. Oropharyngeal profile before (**a**) and after (**b**) the operation. Velopharyngeal profile before (**c**) and after (**d**) the operation.

or persistent narrowing of the velopharynx, which is frequently ignored. The feasibility and increased efficacy of suspending the palatopharyngeal muscle to the pterygomandibular raphe using a knotless bidirectional reabsorbable suture with an external approach has been demonstrated through barbed reposition pharyngoplasty [25]. The author modified the procedure by submucosal interrupted fixation of the palatopharyngeal muscle to the pterygomandibular raphe using a 2-0 Dexon suture. A preliminary report (unpubl. data) shows that the retropalatal space can be further enlarged to a more lateral and anterior position when compared to traditional UPPP (Fig. 2). Oropharyngeal and velopharyngeal profiles from before and after the operation are shown in Figure 3.

Tension-Relieving Dissection

Tonsillar wound dehiscence is a common occurrence in OSA patients after UPPP. The cause can be attributable to high tension induced by suturing of the anterior/posterior pillars and superior pharyngeal constrictor muscle. Tension-relieving dissection (TRD) is performed while a horizontal tension ridge develops after the closure of tonsillar fossa in palatal surgery. The procedure requires a plasma-mediated instrument and is initiated with a longitudinal cut on the mostly tense ridge that usually emerges at the junction of the posterior pillar and posterior pharyngeal wall. Cutting is extended upwards for 1 cm and downwards below the inferior pole of the tonsillar fossa. The incision depth enters the submucosal layer without damage to the posterior pharyngeal wall muscle. The mucosal edge is dissected from the underlying muscles to disperse the tonsillar fossa tension and decrease its dehiscence. In our preliminary data, the dehiscence rate can be significantly reduced from 66.7% (without TRD) to 6.7% (with TRD) (unpubl. data). Postoperative pain and bleeding are also decreased after the addition of tension-relieving dissection in palatal surgery.

Perils and Pitfalls

1. The purposes of palatal surgery, instead of normalization of AHI, are to improve clinical symptoms, decrease severity and reduce major complications and mortality.
2. Palatal surgery, as part of multilevel surgery, can be performed in conjunction with positional therapy and oropharyngeal muscle training to improve the short-term and long-term outcomes.
3. The treatment endpoint for palatal surgery is suggested to include the significant improvement of clinical symptoms, patient satisfaction, AHI <15, and a low risk of major complications revealed by biomarkers.
4. DISE or sleep imaging has gradually replaced traditional wakeful examination in the formation of a surgical plan.
5. The type of palatal surgery can be selected according to the pattern of collapse, i.e., palatoplasty for A-P collapse, pharyngoplasty for lateral collapse.
6. Suspension of the palatopharyngeal muscle to the pterygomandibular raphe seems to be effective in the further enlargement of the retropalatal space.
7. The addition of tension-relieving dissection in palatal surgery is likely to decrease tonsillar wound dehiscence and related postoperative pain and complications.

References

1 Fujita S, Conway W, Zorick F, et al: Surgical correction of anatomic abnormalities in obstructive sleep apnea syndrome: uvulopalatopharyngoplasty. Otolaryngol Head Neck Surg 1981;89:923–934.
2 Sher AE, Schechtman KB, Piccirrillo JF: The efficacy of surgical modifications of the upper airway in adults with obstructive sleep apnea syndrome. Sleep 1996;19:156–177.
3 Li HY, Wang PC, Li LA, Chen NH, Fang TJ: Prediction of uvulopalatopharyngoplasty outcome: anatomy-based staging system versus severity-based staging system. Sleep 2006;29:1537–1541.
4 Li HY, Chen NH, Shu YH, Wang PC: Changes of quality of life and respiratory disturbance after extended uvulopalatal flap surgery in patients with obstructive sleep apnea. Arch Otolaryngol Head Neck Surg 2004;130:195–200.
5 Li HY, Huang YS, Chen NH, Fang TJ, Wang PC: Mood improvement after surgery for obstructive sleep apnea. Laryngoscope 2004;116:1098–1102.
6 Li HY, Lee LA, Yu JF, et al: Changes of snoring sound after relocation pharyngoplasty for obstructive sleep apnoea: the surgery reduces mean intensity in snoring which correlates well to apnoea-hypopnoea index. Clin Otolaryngol 2015;40:98–105.
7 Sullivan CE, Issa FG, Berthon-Jones M, Eves L: Reversal of obstructive sleep apnoea by continuous positive airway pressure applied through the nares. Lancet 1981;1:862–865.
8 Liu SY, Huon LK, Powell NB, et al: Lateral pharyngeal wall tension after maxillomandibular advancement for obstructive sleep apnea is a marker for surgical success: observations from drug-induced sleep endoscopy. J Oral Maxillofac Surg 2015;73:1575–1582.
9 Nagendran M, Carlin AM, Bacal D, et al: Self-reported remission of obstructive sleep apnea following bariatric surgery: cohort study. Surg Obes Relat Dis 2015;11:697–703.

10 Li HY, Wang PC, Hsu CY, Lee SW, Chen NH, Liu SA: Concomitant nasal and palatopharyngeal surgery for obstructive sleep apnea: simultaneous or staged. Acta Otolaryngol 2005;125:298–303.
11 Li HY, Wang PC, Hsu CY, Chen NH, Lee LA, Fang TJ: Same-stage palatopharyngeal and hypopharyngeal surgery for severe obstructive sleep apnea. Acta Otolaryngol 2004;124:820–826.
12 Lin HC, Friedman M, Chang HW, Gurpinar B: The efficacy of multi-level surgery of the upper airway in adults with obstructive sleep apnea/hypopnea syndrome. Laryngoscope 2008;18:902–908.
13 Li HY, Cheng WN, Chuang LP, et al: Positional dependency and surgical success of relocation pharyngoplasty among patients with severe obstructive sleep apnea. Otolaryngol Head Neck Surg 2013;149:506–512.
14 Ieto V, Kayamori F, Montes MI, et al: Effects of oropharyngeal exercises on snoring – a randomized trial. Chest 2015;148:683–691.
15 Lee LA, Huang CG, Chen NH, Wang CL, Fang TJ, Li HY: Severity of obstructive sleep apnea syndrome and high-sensitivity C-reactive protein reduced after relocation pharyngoplasty. Otolaryngol Head Neck Surg 2011;144:632–638.
16 Kezirian EJ, Hohenhorst W, de Vries N: Drug-induced sleep endoscopy: the VOTE classification. Eur Arch Otolaryngol 2011;268:1233–1236.
17 Chuang LP, Chen NH, Li HY, et al: Dynamic upper airway changes during sleep in patients with obstructive sleep apnea syndrome. Acta Otolaryngol 2009;129:1474–1479.
18 Powell N, Riley R, Guilleminault C, et al: A reversible uvulopalatal flap for snoring and sleep apnea syndrome. Sleep 1996;19:593–599.
19 Fairbanks DNF: Uvulopalatopharyngoplasty techniques, pitfalls, and risk management; in Fairbanks DNF, Mickelson SA, Woodson BT (eds): Snoring and Obstructive Sleep Apnea, ed 3. Philadelphia, Lippincott Williams & Wilkins, 2003, pp 107–120.
20 Li HY, Li KK, Chen NH, Wang PC: Modified uvulopalatopharyngoplasty: the extended uvulopalatal flap. Am J Otolaryngol 2003;24:311–316.
21 Friedman M, Ibrahim H, Vidyasagar R, et al: Z-palatoplasty (ZPP): a technique for patients without tonsils. Otolaryngol Head Neck Surg 2004;131:89–100.
22 Cahali MB: Lateral pharyngoplasty: a new treatment for obstructive sleep apneahypopnea syndrome. Laryngoscope 2003;113:1961–1968.
23 Pang KP, Woodson BT: Expansion sphincter pharyngoplasty: a new technique for the treatment of obstructive sleep apnea. Otolaryngol Head Neck Surg 2007;137:110–114.
24 Li HY, Lee LA: Relocation pharyngoplasty for obstructive sleep apnea. Laryngoscope 2009;119:2472–2477.
25 Vicini C, Hendawy E, Campanini A, et al: Barbed reposition pharyngoplasty (BRP) for OSAHS: a feasibility, safety, efficacy and teachability pilot study. "We are on the giant's shoulder". Eur Arch Otorhinolaryngol 2015;272:3065–3070.
26 Friedman M, Ibrahim H, Lowenthal S, et al: Uvulopalatoplasty (UP2): a modified technique for selected patients. Laryngoscope 2004;114:441–449.
27 Woodson BT, Toohill RJ: Transpalatal advancement pharyngoplasty for obstructive sleep apnea. Laryngoscope 1993;103:169–176.
28 Li HY, Lee LA, Fang TJ, Lin WY: Evaluation of velopharyngeal function after relocation pharyngoplasty for obstructive sleep apnea. Laryngoscope 2010; 120:1069–1073.

Hsueh-Yu Li, MD, FACS
Department of Otolaryngology, Chang Gung Memorial Hospital
5, Fushing Street
Gueishan Shiang, Taoyuan 33305, Taiwan (ROC)
E-Mail hyli38@cgmh.org.tw

Lin H-C (ed): Sleep-Related Breathing Disorders. Adv Otorhinolaryngol. Basel, Karger, 2017, vol 80, pp 81–89
DOI: 10.1159/000470871

Updated Hypopharyngeal Surgery for Sleep Apnea

B. Tucker Woodson

Department of Otolaryngology, Medical College of Wisconsin, Milwaukee, WI, USA

Abstract

The failure of traditional upper airway surgery such as uvulopalatopharyngoplasty has been attributed to lower airway obstruction related to base of tongue collapse. Multiple procedures, including glossectomy, tongue base radiofrequency, genioglossus advancement, and tongue suspension techniques, have been advocated to improve success rates. No consensus exists on which subsets of patients are best treated by individual approaches and direct comparative data are lacking. The selection of procedures must be based on individual patient needs and the relative potential benefits and risks.

The first surgeries for obstructive sleep apnea (OSA) were palatal procedures. Fujita et al. [1] described uvulopalatopharyngoplasty (UPPP) as a procedure to remove tonsils and tissues of the palate and pharynx to treat OSA. Although the procedure frequently reduced the severity, significant apnea often remained. With the advent of nasal continuous positive airway pressure, the procedure is infrequently used as the first-line treatment for all but a select group of patients. Attempts to improve palate surgery by more aggressive surgical resection of the palate resulted in higher complications but not greater success rates [2]. Instead, improved outcomes were observed by expanding surgical treatment to nonpalatal sites of airway obstruction. Reducing airway collapse at multiple sites, especially those directed at the hypopharynx and tongue base, have demonstrated higher success rates than palate surgeries alone [3].

Multiple sites have been observed to be either primary or secondary sites of airway obstruction [4–6]. To that end, multiple hypopharyngeal airway surgical approaches have been developed to improve surgical outcomes by treating these non-palatal sites. Some procedures conceptually attempt to increase airway volume while others aim to alter tissue or airway wall compliance. Procedures may be either skeletal, with the goal of advancing a segment of the facial skeletal framework, or soft tissue, with the goal of tissue stiffening or volume removal. Historically, aside from a tracheotomy that bypasses the airway (and often has unacceptable morbidity), maxillomandibular advancement has demonstrated the highest success rates [7]. More limited and site-specific procedures have been described and include segmental mandibular advancement osteotomies, soft tissue removal (glossectomy and lingual tonsillectomy), and methods of tongue suspension. Recently, cranial nerve stimulation was developed to assist treating complex patterns of upper airway collapse. Unfortunately, no consensus exists on which, if any, procedures are best for given patients. The enormous variability of procedures is beyond the scope of this discussion. However, the anatomy, the concepts of surgery, and selected procedures will be presented here.

The selection of any treatment in medicine results from balancing the benefits, harms, patient/physician preferences, and cost. This can be formally addressed using "grading" systems [8]. Although many articles present data supporting the addition of the tongue base and multilevel procedures over palate surgery alone, many address only short-term surgical benefits. Sparse data are available on surgical risk, long-term benefit, and postoperative morbidity. As a result, treatment decisions are often difficult and controversial, and although most surgeons agree that surgical outcomes are improved with hypopharyngeal procedures, few agree which are best for any given patient. Ultimately, it is the balance between patient anatomy, procedure availability, published procedure outcomes, and procedure morbidity that determines patient and procedure selection.

Anatomy

The mechanisms of how the pharyngeal airway contributes to sleep-related upper airway collapse is complex and incompletely understood. It is agreed that the upper airway in most OSA patients is smaller and more vulnerable to collapse than in those without OSA. Since the lower pharynx is usually not the primary site of obstruction, other mechanisms other than strictly size and primary collapse are likely to be involved and, ultimately, an understanding of the mechanisms of disease will probably lead to more successful surgical techniques. Although the pharyngeal airway is considered the primary route of breathing, respiration may be either nasal or oral. Since both oral and pharyngeal routes intersect in the lower pharynx, hypopharyngeal surgeries may affect both. Beneficial mechanisms of lower pharyngeal surgery may also include reducing total airway resistance, altering a primary site of obstruction, preventing the mass effect of the tongue collapsing against the soft palate, or improving the dynamics of the oropalatal airway and oral respiration, or other yet to be determined effects.

Surgical Algorithms

Hypopharyngeal surgery is usually applied as part of a multilevel approach to OSA surgery. In any single individual treatment of the nose, the nasopharynx, oropharynx, and hypopharynx may be required. Aside from a tracheotomy, which obviously bypasses the upper airway, no single technique uniquely addresses all sites. Maxillomandibular advancement procedures may be successful single-stage procedures, but acceptance is low for many patients with normal craniofacial esthetics and mild or moderate disease, despite high success rates having been published. To address different patterns of airway obstruction, various algorithms or "recipes" have been developed. The 2-phase "Stanford" treatment algorithm described by Powell and Riley [9] that utilized UPPP with or without genioglossus advancement followed by maxillomandibular advancement is an example [8]. With the development of other multilevel procedures, a wide variety of other combinations have also been described that include glossectomy, hyoid suspension, tongue suspension, radiofrequency ablation, lingual tonsillectomy, and others. Unfortunately, few comparisons exist to determine what is best for a given patient. Such heterogeneity clearly advocates for more and better data to grade surgical treatment decisions. There is a need not only for studies demonstrating the short-term benefit, but also of harm and long-term outcomes.

An additional contributor to surgical decision making is the assessment of perioperative risk and recovery. Some multilevel procedures are best performed as individual distinct surgical procedures while others may be combined. Combined procedures are generally associated with a greater postoperative risk. Recent studies support that in milder OSA less aggressive procedures (such as tongue base radiofrequency) can be combined safely [10–12].

Airway Assessment

Airway collapse is the result of a vulnerable and small upper airway. Risks result from the interaction of a craniofacial structure, soft tissue size, and obesity [13, 14]. Ultimately, a constellation of these abnormalities causes OSA. In nonobese patients, skeletal structural abnormalities are often a major contributor, while in obese patients both craniofacial and soft tissues contribute. Although surgical therapies are often directed at the primary contributor, with skeletal procedures directed at those with skeletal deficiencies and soft tissue procedures directed at those with excessive soft tissues, such an approach is not uniform.

Multiple methods have been used to assess the airway and assist in selecting surgical techniques. Methods include a general clinical exam, sitting and supine endoscopy, sedated endoscopy, and radiologic studies. No method is uniformly accepted and each may contribute useful information. A general physical exam using the Fujita classification that stratifies individuals into type I (upper pharynx), type III (lower pharynx), and type II (combined upper and lower pharynx narrowing or obstruction) has now been replaced by drug-induced sedated endoscopy (DISE) and airway phenotyping [15, 16]. The landmark Friedman staging system, which has demonstrated a predictive value for UPPP, may be useful in selecting patients who might benefit from hypopharyngeal surgery but is of uncertain benefit in selecting which hypopharyngeal surgery to perform [17].

Drug-Induced Sedated Endoscopy

The Müller maneuver artificially creates a dynamic collapse to try to measure the airway. As an alternative, multiple authors have advocated assessing the upper airway during sedation. Conceptually, DISE might provide better information than examinations during wakefulness. The VOTE classification is widely used to provide qualitative and quantitative insight into patterns of airway collapse. This scheme evaluates collapse at 3 segments (the velopharynx, oropharynx, and tongue) and 1 structure (the epiglottis) to better describe obstruction and improve surgical outcomes [15]. However, despite its conceptual potential, VOTE has yet to demonstrate the predictive success of hypopharyngeal surgical procedures [18, 19].

Phenotypes

Complex airway characteristics determine airflow and airway resistance. Features include sagittal (longitudinal) shape, cross-sectional size, vertical length, airway centric curvature, and contour. Condensing these characteristics into a single descriptive point of collapse for each segment (velopharynx, oropharynx, hypopharynx, etc.) is most probably inadequate. However, creating a method to detail descriptions of multiple meaningful characteristics of the airway lumen is a challenge. A different approach that does not only assess airway collapse has been described. This method describes upper airway phenotypes by describing the airway lumen in relation to "key" surrounding tissue structures [16]. Invariably, critical airway characteristics relate to key structures. For example, in the hypopharynx, the anterior pharyngeal wall is the tongue and it is easily observable that tongue anatomy differs among individuals with OSA. Moore and Phillips [20] developed a classification that was applied to lateral cephalometric X-rays. This defined the hypopharyngeal anatomy as 3 basic types (A, B, C). Type A described proximal tongue-base obstruction, types B1 and B2 described proximal tongue-base obstruction combined with either distal tongue base or retroepiglottic obstruction, and type C described retroepiglottic obstruction alone. This has been modified and applied to endoscopic evaluation.

The endoscopic method divides the anterior wall of the hypopharynx and posterior tongue into several key elements. These include the upper hypopharynx/tongue (above the tip of the epiglottis), the lower hypopharynx/tongue (below the tip of the epiglottis), the lingual tonsils, and epiglottis. The first 2 are landmarks to assess the airway lumen (upper and lower tongue), and clinical experience has observed that the second 2 are surgically important structures that may be the independent cause of airway obstruction. In using these structures, a "modified Moore" classification is described.

Cephalometrics
The primary metric used on cephalometric X-rays to identify patients for treatment was the posterior airway space. This is defined differently by authors either as the narrowest anterior posterior distance in the tongue base segment, the width of the airway on a line parallel to the occlusal plane, or the width along a line extending from the B point to the gonion (the most posterior inferior point of the mandible). A distance of less than 11 mm is considered abnormal. Data suggest that the vertical length of the airway measured by the mandibular plane to hyoid distance (MP-H) is useful in predicting surgical outcomes. Lengths greater than 22 mm are associated with lower success rates.

Base of the Tongue Techniques

A wide variety of hypopharyngeal techniques have been described. Procedures include ablation glossectomy, excisional partial glossectomy, limited mandibular osteotomies, mandibular advancement, hyoid suspension, and tongue suspension [21–24].

Glossectomy
Glossectomy is a procedure that reduces the volume of the tongue by the direct removal of tissues. Multiple techniques and tools for glossectomy have been described. These include midline glossectomy, submucosal glossectomy (SMILE), lingualplasty, anterior glossectomy, transoral robotic glossectomy (TORS), radiofrequency ablation, and others [25–28]. The outcomes of glossectomy for OSA are almost universally reported in combination with UPPP without control groups, making their assessment difficult. In a systematic review of glossectomy, Murphey et al. [29] observed a mean change in the apnea-hypopnea index of 27.8 (95% CI 22.6–33.0) and mean change in Epworth score of 5.5 (95% CI 3.3–7.2). Long-term outcomes are uncertain, with data suggesting significant long-term relapse rates [23]. Complications and side effects are inconsistently presented, making grading assessments difficult.

Indications/Contraindications
Glossectomy may impair speech, swallowing, and taste. The impairment is likely correlated to the

aggressiveness of muscle resection. Data with lingual tonsillectomy, which most likely has similar risks, suggests that the risk, severity, and duration of taste changes are variable [30]. Those patients in whom even minor changes in taste would cause a potential disability should be excluded. Patients with existing swallowing or speech problems, untreated severe extraesophageal reflux (which would likely impair healing and recuperation), or those with COPD where the development of swallowing difficulties poses a major medical risk, are best avoided unless alternative therapies are not available. Patients with overt craniofacial abnormalities may be treated with maxillofacial surgery.

Tissue volume reduction may be performed using direct excision or ablation radiofrequency of the tongue base (radiofrequency tongue ablation, or RFTA). Glossectomy modifies the shape and size of the soft tissue tongue base and the hypopharynx. Airway space is increased by reducing the bulk of soft tissue. Although conceptually simple, using traditional methods, morbidity of the procedure may be high and exposure difficult. Glossectomy morbidity and complications likely correlate to the aggressiveness of resection and the type of the surgical tools used. The removal or damage of increasingly more tissue worsens the perioperative recovery (pain, edema, dysphagia, bleeding, infection). Tools, such as a laser and cautery, with significant thermal tissue effects are associated with more pain than lower-temperature devices that use plasma (such as Coblation; Smith Nephew, Austin, TX, USA) [30]. Risks of airway obstruction, bleeding, infection, dehydration, and dysphagia have been reported in up to 25% of patients [31]. The relatively high morbidity limits the population of patients who are candidates. For this reason, RFTA has been used to reduce tongue volume. The advantages of RFTA include a low surgical morbidity and research studies providing high-level evidence of effect [32]. The disadvantages include the potential requirement for multiple treatment sessions and a limit to the location and volume of tissue that may be reduced. The limitation in tissue treatment volumes may limit the clinical "cure" and may explain why effectiveness is partial and residual OSA is often present after treatment. Recently, alternative algorithms for RFTA have been presented [33, 34]. These treat multiple sites in the tongue and not just the posterior third, as was initially described. RFTA complications include tongue abscess, infection (cellulitis), tongue weakness, changes in speech and swallowing, and acute airway obstruction and edema. If future studies demonstrate that the medical risk of mild residual apnea is low, RFTA may have an increased role is OSA treatment.

Tongue volume excision may be directly performed using multiple techniques and tools. Tools include cold excision, laser, diathermy, and bipolar and monopolar radiofrequency. Systematic reviews support that, when combined with UPPP, a significant short-term reduction in disease severity is observed [29]. Long-term results are not well evaluated but the potential for significant relapse exists [35]. Midline posterior glossectomy and lingualplasty were initially performed with a CO_2 laser. These partial glossectomies were directed at enlarging the lower pharyngeal airway. CO_2 laser lingualplasty following UPPP failure reduced the apnea-hypopnea index to <20 in 70% of patients with short-term follow-up. Complication rates were high, approaching 25%.

Plasma tools (Coblation; Smith Nephew) and newer technologies have been described for both glossectomy and lingual tonsillectomy with lower complication rates. The procedure can be performed as outpatient surgery in select patients. Open techniques with and without fiber-optic endoscopes have been described [36–38]. A submucosal approach has also been used, termed submucosal minimally invasive lingual excision (SMILE) [39]. The author prefers the open approach due to concerns about a closed space seroma, hematoma, or abscess. The open approach uses a midline traction suture for initial exposure. A midline incision placed anterior to the circumvallate papillae begins the procedure, which is then followed by further exposure of the tongue dorsum using 2 lateral traction sutures. The procedure is performed using angled rigid fiber-optic endoscopes.

Damage to the neurovascuolar bundle is a risk of the procedure. The lingual artery may be used to locate the neurovascular pedicle of the tongue. A knowledge of lingual anatomy is essential. To reduce the risk of damaging the lingual artery, ultrasound/Doppler may be used to map and mark the location of the anterior segment of the lingual artery and the location of the profunda segment on the surface of the tongue. Once these are identified, the "safe area" for resection between the oblique segments, which lie in the posterior lateral tongue, are known. Alternatively, an approach without ultrasound or Doppler may be performed by using a more conservative resection limited to the posterior third of the tongue and staying within 1 cm of the midline. A suprahyoid branch of the lingual artery may be encountered with bleeding that might require suction electrocautery. After removal of the midline segment, prolapse of tongue tissues generally obliterates the space created. Sutures decrease dead space and reduce the formation of an excessive furrow in the midline of the tongue. The posterior wound is open to prevent abscess formation. Both broad-spectrum (aerobic and anaerobic antibiotics) and treatment gastroesophageal reflux is recommended.

A slightly modified glossectomy using Coblation (Smith Nephew) has been described that removes more lateral tissue (with a similarity to lingualplasty) [38]. TORS has also been described as a method to perform glossectomy [40]. All techniques seem to follow the general rule that "the more aggressive the resection, the higher the success rate … the higher the complication rate." TORS UPPP glossectomy studies correlate higher success rates with a volume reduction of approximately 10 cm^3. Lesser or greater volume reductions perform more poorly.

Lingual tonsillar hypertrophy may cause or contribute to sleep apnea, and removal may have a marked effect on sleep apnea. In OSA patients, lingual tonsillectomy can be challenging, with exposure difficult or impossible. A tracheotomy was required for some patients. Newer surgical techniques combining endoscopes and excision (similar to glossectomy, as outlined above) or using TORS allow easier removal of lingual tonsil tissue.

Genioglossus Advancement

Limited surgical osteotomies of the anterior mandible were among the first described surgical procedures for the treatment of OSA. Conceptually, the procedure reconstructs the airway by advancing the genioglossus muscle insertion. This may physically create a volume of tissue and/or increase tension on the genioglossus muscle to increase the vector of force the muscle creates to maintain patency. A wide variety of osteotomy techniques have been used from genioplasty to a rectangular osteotomy preserving inferior mandibular integrity. As an isolated procedure, no studies have evaluated its efficacy, but combined with UPPP, multiple case series have shown a 60% effectiveness. Side effects include the risk of muscle relapse, partial teeth numbness, and the rare risk of jaw fracture. The outcomes from a systemic review are presented in Table 1 [41].

Genioglossus Advancement with Suspension

An alternative to moving the muscle insertion is to intrinsically support the genioglossus muscle without bone movement. Support of the soft tissues has been described using fascia, suture, or wire. These techniques do not create tension in the muscle but translate the force to an implant. Recently, techniques have been described using suture to support the tongue using special construction bone screws in a submental approach (Airvance; Medtronic ENT, Jacksonville, FL,

Table 1. Overview of systematic reviews for hypopharyngeal surgery [41]

Procedure	Studies included, *n*	Patients, *n*	AHI baseline/final	% success	Evidence level
Glossectomy + UPPP	18	498	48.1/19.1	59.6	C
GGA + UPPP	7	151	53.2/23.8	61.1	C
GGA + HS + UPPP	12	467	54.6/22.7	61.1	C
TS + UPPP	8	167	44.0/16.6	62.3	C
TS alone	8	82	32.0/18.8	36.6	C
HS alone	9	101	37.3/23.0	Not available	C
Radiofrequency (multilevel)	11	Not available	40% reduction from baseline	Not available	B

AHI, apnea-hypopnea index, GGA, genioglossus advancement; UPPP, uvulopalatopharyngoplasty; HS, hyoid suspension; TS, tongue suspension.

USA; Siesta Medical, Sunnyvale, CA, USA). Initial implants were performed transorally, but newer methods use a less invasive submental approach [42]. Several case series support the techniques and a single small study compares suspension to osteotomy [43]. Suspension has the distinct advantage of being less invasive and being potentially reversible. Data on the benefit compared to UPPP alone are lacking, but significant data support that the procedures can be performed with low morbidity [44]. The outcomes from a systemic review are detailed in Table 1 [41].

Hyoid Suspension

In animals, the hyoid bone is attached to the skull and strongly supports the pharynx, while in humans is free floating in the neck. Pulling the hyoid forward using suspension to the mandible as well as suspension to the more anteriorly situated thyroid cartilage has been described. Hyoid movement may stabilize the hypopharyngeal airway, including the epiglottis. The hyoid may be suspended anteriorly and superiorly to the mandible or anteriorly and inferiorly to the thyroid cartilage [45]. Hyoid myotomy may be performed with other surgeries, such as UPPP or genioglossal advancement. Controlled studies are lacking. The ideal candidate to benefit is undetermined; however, those who have a favorable mandibular plane to hyoid distance (<25 mm) may be the most likely to benefit. In addition to the anterior placement of the hyoid, moving the hyoid may tense the lateral pharyngeal walls and augment the effects of tracheal tug. The outcomes from a systemic review are described in Table 1 [46].

Epiglottidectomy

Removal of the epiglottis has been performed with both midline glossectomy and lingualplasty. In both adults and children, removal of an abnormal epiglottis has improved OSA and upper airway obstruction. Partial resection of the epiglottis with a CO_2 laser has been described as a safe procedure in adults.

Multilevel Radiofrequency

RFTA has been included as part of multiple algorithms for OSA. Many include RFTA with UPPP. RFTA alone has been assessed separately and data supports a significant but partial effect. Data also reveal a low morbidity for the procedure [47].

Conclusion

We have provided an overview of hypopharyngeal surgery. Although commonly performed, uncertainty exists regarding the best surgical options and algorithms for patients. Additional data will provide insight into which procedures offer the most benefit and the least harm. Additional studies will also provide insight into simultaneous or staged procedures for selected algorithms.

References

1 Fujita S, Conway W, Zorick F, et al: Surgical correction of anatomic abnormalities of obstructive sleep apnea syndrome: uvulopalatopharyngoplasty. Otolaryngol Head Neck Surg 1981;89: 923–934.

2 Zohar Y, Finkelstein Y, Strauss M, Shivilli Y: Surgical treatment of obstructive sleep apnea: technical variations. Arch Otolaryngol Head Neck Surg 1993;119: 1023–1029.

3 Caples SM, Rowley JA, Prinsell JR, Pallanch JF, Elamin MB, Katz SG, Harwick JD: Surgical modifications of the upper airway for obstructive sleep apnea in adults: a systematic review and meta-analysis. Sleep 2010;33:1396–1407.

4 Morrison DL, Launois SH, Isono S, et al: Pharyngeal narrowing and closing pressures in patients with obstructive sleep apnea. Am Rev Respir Dis 1993;148: 606–611.

5 Woodson BT, Wooten MR: Manometric and endoscopic localization of airway obstruction after uvulopalatopharyngoplasty. Otolaryngol Head Neck Surg 1994;111:38–43.

6 Katsantonis GP, Moss K, Miyazaki S, et al: Determining the site of airway collapse in obstructive sleep apnea with airway pressure monitoring. Laryngoscope 1993;103:1126–1131.

7 Won CH, Li KK, Guilleminault C: Surgical treatment of obstructive sleep apnea: upper airway and maxillomandibular surgery. Proc Am Thorac Soc 2008;5: 193–199.

8 Aurora RN, Casey K, Kristo D, et al: Practice parameters for the surgical modifications of the upper airway for obstructive sleep apnea in adults. Sleep 2010;33:1408–1413.

9 Powell N, Riley R: A surgical protocol for sleep disordered breathing. Oral Maxillofac Surg Clin North Am 1995;7: 345–356.

10 Mickelson SA, Hakim I: Is postoperative intensive care monitoring necessary after uvulopalatopharyngoplasty? Otolaryngol Head Neck Surg 1998;119:352–356.

11 Pang KP, Siow JK, Tseng P: Safety of multilevel surgery in obstructive sleep apnea: a review of 487 cases. Arch Otolaryngol Head Neck Surg 2012;138:353–357.

12 Kandasamy T, Wright ED, Fuller J, Rotenberg BW: The incidence of early post-operative complications following uvulopalatopharyngoplasty: identification of predictive risk factors. J Otolaryngol Head Neck Surg 2013;42:15.

13 Tsuiki S, Isono S, Ishikawa T, Yamashiro Y, Tatsumi T, Nishino T: Anatomical balance of the upper airway and obstructive sleep apnea. Anesthesiology 2008;108:1009–1015.

14 Dempsy JA, Skatrud JB, Jacques AJ, Ewanowski SJ, Woodson T, Hanson PR, Goodman B, Young T: Anatomical determinates of sleep disordered breathing across the spectrum of clinical and nonclinical subjects. Chest 2002;122:40–51.

15 Kezirian EJ, Hohenhorst W, de Vries N: Drug-induced sleep endoscopy: the VOTE classification. Eur Arch Otorhinolaryngol 2011;268:1233–1236.

16 Woodson BT: A method to describe the pharyngeal airway. Laryngoscope 2015; 125:1233–1238.

17 Friedman M, Ibrahim H, Bass L: Clinical staging for sleep-disordered breathing. Otolaryngol Head Neck Surg 2002;127: 13–21.

18 Eichler C, Sommer JU, Stuck BA, Hörmann K, Maurer JT: Does drug-induced sleep endoscopy change the treatment concept of patients with snoring and obstructive sleep apnea? Sleep Breath 2013;17:63–68.

19 Golbin D, Musgrave B, Succar E, Yaremchuk K: Clinical analysis of drug-induced sleep endoscopy for the OSA patient. Laryngoscope 2016;126:249–253.

20 Moore KE, Phillips C: A practical method for describing patterns of tongue-base narrowing (modification of Fujita) in awake adult patients with obstructive sleep apnea. J Oral Maxillofac Surg 2002; 60:252–260.

21 Riley RW, Powell NB, Guilleminault C: Obstructive sleep apnea syndrome: a review of 306 consecutively treated surgical patients. Otolaryngol Head Neck Surg 1993;108:117–125.

22 Fujita S, Woodson BT, Clark JL, Wittig R: Laser midline glossectomy as a treatment for obstructive sleep apnea. Laryngoscope 1991;101:805–809.

23 DeRowe A, Gunter E, Fibbi A, Lehtimaki K, Vahatalo K, Maurer J, Ophir D: Tongue-base suspension with a soft tissue-to-bone anchor for obstructive sleep apnea: preliminary clinical results of a new minimally invasive technique. Otolaryn Head Neck Surg 2000;122:101–103.

24 Powell NB, Riley RW, Guilleminault C: Radiofrequency tongue base reduction in sleep-disordered breathing: a pilot study. Otolaryngol Head Neck Surg 1999;120:656–664.

25 Mickelson S, Rosenthal L: Midline glossectomy and epiglottidectomy for obstructive sleep apnea syndrome. Larynogoscope 1997;107:614–619.

26 Woodson BT, Fujita S: Clinical experience with lingualplasty as part of the treatment of severe obstructive sleep apnea. Otolaryngol Head Neck Surg 1992;107:40–48.

27 Friedman M, Soans R, Gurpinar B, Lin H, Joseph N: Evaluation of submucosal minimally invasive lingual excision technique for treatment of obstructive sleep apnea/hypopnea syndrome. Otolaryngol Head Neck Surg 2008;139:378–384.

28 Vicini C, Dallan I, Canzi P, Frassineti S, Nacci A, Seccia V, Panicucci E, Grazia La Pietra M, Montevecchi F, Tschabitscher M: Transoral robotic surgery of the tongue base in obstructive sleep apnea-hypopnea syndrome: anatomic considerations and clinical experience. Head Neck 2012;34:15–22.

29 Murphey AW, Kandl JA Nguyen SA, Weber AC, Gillespie MB: The effect of glossectomy for obstructive sleep apnea: a systematic review and meta-analysis. Otolaryngol Head Neck Surg 2015;153: 334–342.

30 Robinson S, Lewis R, Norton A, McPeake S: Ultrasound-guided radiofrequency submucosal tongue-base excision for sleep apnoea: a preliminary report. Clin Otolaryngol Allied Sci 2003; 28:341–345.

31 Glazer TA, Hoff PT, Spector ME: Transoral robotic surgery for obstructive sleep apnea: perioperative management and postoperative complications. JAMA Otolaryngol Head Neck Surg 2014;140: 1207–1212.

32 Woodson BT, Steward DL, Weaver EM, et al: A randomized trial of temperature-controlled radiofrequency, continuous positive airway pressure, and placebo for obstructive sleep apnea syndrome. Otolaryngol Head Neck Surg 2003;128: 848–861.

33 MacKay SG, Carney AS, Woods C, Antic N, McEvoy RD, et al: Uvulopalatopharyngoplasty and coblation channeling of the tongue for obstructive sleep apnea: a multi-centre Australian trial. J Clin Sleep Med 2013;9:117–124.

34 Hou T, Hu S, Jiang X: Tongue coblation via the ventral approach for obstructive sleep apnea-hypopnea syndrome surgery. Laryngoscope 2012;122:2582–2586.

35 Hou J, Yan Y, Wang B, Zhu K, Sheng Y, et al: Treatment of obstructive sleep apnea-hypopnea syndrome with combined uvulopalatopharyngoplasty and midline glossectomy: outcomes from a 5-year study. Respir Care 2012;57:2104–2110.

36 Woodson BT, Laohasiriwong S: Endoscopic midline glossectomy for obstructive sleep apnea; in Pang KP, Rotenberg BW, Woodson BT (eds): Advanced Surgical Techniques in Snoring and Obstructive Sleep Apnea. San Diego, Plural Publishing, 2013.

37 Robinson S, Ettema SL, Brusky L, Woodson BT: Lingual tonsillectomy using bipolar radiofrequency plasma excision. Otolaryngol Head Neck Surg 2006;134:328–330.

38 Li HY, Lee LA, Kezirian EJ: Efficacy of coblation endoscopic lingual lightening in multilevel surgery for obstructive sleep apnea. JAMA Otolaryngol Head Neck Surg 2016;142:438–443.

39 Maturo SC, Mair EA: Submucosal minimally invasive lingual excision: an effective, novel surgery for pediatric tongue base reduction. Ann Otol Rhin Laryngol 2006;115:624–630.

40 Meccariello G, Cammaroto G, Montevecchi F, Hoff PT, Spector ME, Negm H, Shams M, Bellini C, Zeccardo E, Vicini C: Transoral robotic surgery for the management of obstructive sleep apnea: a systematic review and meta-analysis. Eur Arch Otorhinolaryngol 2017;274: 647–653.

41 Handler E, Hamans E, Goldberg AN, Mickelson S: Tongue suspension: an evidence-based review and comparison to hypopharyngeal surgery for OSA. Laryngoscope 2014;124:329–336.

42 Huang T, Su H, Wang C, Cheng P: Transsubmental tongue-base suspension in treating patients with severe obstructive sleep apnoea after failed uvulopalatopharyngoplasty: our experience. Clin Otolaryngol 2014;39:114–126.

43 Thomas AJ, Chavoya M, Terris DJ: Preliminary findings from a prospective, randomized trial of two tongue-base surgeries for sleep-disordered breathing. Otolaryngol Head Neck Surg 2003;129: 539–546.

44 Handler E, Hamans E, Goldberg AN, Mickelson S: Tongue suspension: an evidence-based review and comparison to hypopharyngeal surgery for OSA. Laryngoscope 2014;124:329–336.

45 Gillespie MB, Ayers CM, Nguyen SA, Abidin MR: Outcomes of hyoid myotomy and suspension using a mandibular screw suspension system. Otolaryngol Head Neck Surg 2011;144:225–229.

46 Song SA, Wei JM, Buttram J, Tolisano AM, Chang ET, Liu SY, Certal V, Camacho M: Hyoid surgery alone for obstructive sleep apnea: a systematic review and meta-analysis. Laryngoscope 2016;126: 1702–1708.

47 Baba RW, Mohan A, Ramesh Metta VVS, Mador J: Temperature controlled radiofrequency ablation at different sites for treatment of obstructive sleep apnea syndrome: a systematic review and meta-analysis. Sleep Breath 2015;19:891–910.

Prof. B. Tucker Woodson, MD
Department of Otolaryngology, Medical College Wisconsin
9200 W Wisconsin Avenue
Milwaukee, WI 53226 (USA)
E-Mail bwoodson@mcw.edu

Lin H-C (ed): Sleep-Related Breathing Disorders. Adv Otorhinolaryngol. Basel, Karger, 2017, vol 80, pp 90–98
DOI: 10.1159/000470872

Updated Minimally Invasive Surgery for Sleep-Related Breathing Disorders

Bhik Kotecha

Royal National Throat, Nose and Ear Hospital (UCLH), London, UK

Abstract

Sleep-related breathing disorders (SRBD) consist of a variety of clinical entities, ranging from primary snoring to severe obstructive sleep apnoea at the opposite extreme. Although it has no official definition, minimally invasive surgery infers surgery where the procedure is clinically less aggressive and may involve a much smaller incision or none at all. It also implies that such procedures may result in a quicker recovery and reduced morbidity, and that they are often conducted under local anaesthesia as day cases. Minimally invasive surgery in SRBD may be performed on its own or may be conducted in conjunction with other surgical procedures or in association with non-surgical treatment modalities such as oral appliances. Indeed, it may be considered in patients experiencing difficulty tolerating treatment with continuous positive airway pressure. This is with a view to improving the upper airway dimensions and thus reducing the pressure requirement with this type of treatment modality, hence improving its compliance and efficacy. Here we address the role of minimally invasive surgery in SRBD within various anatomical regions of the upper airway, namely the nose, soft palate, and the tongue.

Sleep-related breathing disorders (SRBD) consist of a variety of clinical entities, ranging from primary snoring to severe obstructive sleep apnoea (OSA) at opposite extremes. The management of SRBD attracts a multidisciplinary approach within which surgery has a definitive role to play [1]. In order to attain a successful outcome from surgical intervention, it is crucial to adequately evaluate and subsequently select the patient for the appropriate procedure. Otolaryngologists are well positioned in this field and have the ability to assess and rectify upper airway obstructions [2].

Ambulatory home sleep studies or hospital-based polysomnography could assist in identifying where the patient is positioned in the spectrum of SRBD. In general, it is accepted that patients with moderate or severe OSA should be offered continuous positive airway pressure (CPAP) as the treatment of choice, but the compliance and adherence with this form of therapy is not optimum, and hence the role of surgery becomes useful in either facilitating CPAP therapy or offering an alternative management option [1, 3, 4].

A number of different techniques have been utilised to evaluate upper airway obstruction and include a variety of imaging modalities, acoustic analysis, pressure manometry, and endoscopy. Of these, drug-induced sedation endoscopy (DISE) and pressure manometry seem to be more practical and effective in identifying levels of dynamic anatomical obstruction of the upper airway during sleep, thus allowing site-specific treatment [5]. A number of controversies have been linked with DISE, but most have been addressed by numerous studies specifically targeting issues with the depth of sedation, inter-rater reliability, and awake versus asleep assessment [6–10].

Surgical intervention in any field is potentially associated with morbidity, and surgery for SRBD is no exception. However, minimally invasive surgery sounds more acceptable as it tends to infer that this kind of procedure is less radical than standard procedures, could be performed under local anaesthesia as day cases, and perhaps the incidence of complications and morbidity encountered may be reduced. In the field of SRBD surgery, there is no official or formal definition of minimally invasive surgery and new techniques may be described as minimally invasive when first introduced, such as laser-assisted uvulopalatoplasty, only later to be associated with far more increased morbidity and complications [11–13]. Palatal surgery is probably the most common surgery performed in this group of patients, and for the soft palate one could propose that minimally invasive surgery is such that the procedure does not involve any excision or change in the shape of the soft palate. As such, this would then include injecting sclerosing agents to stiffen the soft palate, inserting implants or performing interstitial radiofrequency thermal ablation to alter the elasticity of the soft palate as opposed to changing its shape. Similarly, minimally invasive surgery for the nose and the base of the tongue may include radiofrequency application to alter the upper airway dynamics.

Nasal Surgery

Nasal obstruction or congestion is common and causes mouth breathing which in turn destabilises the upper airway and exacerbates SRBD in some patients [14]. Problems within the nose can be anatomical (e.g., nasal valve collapse or deviated septum) or pathophysiological (e.g., rhinosinusitis or nasal polyposis). Although some rhinologists may consider nasal valve surgery or septoplasty to be relatively "minimally invasive," for the purpose of this discussion only surgery for the reduction of inferior turbinates will be treated as such. Various studies have reported that chronic nasal congestion can cause snoring and result in daytime sleepiness [15–18]. Compromised nasal airflow causes increased resistance and, therefore, more turbulent flow is encountered, which in turn can lead to an open mouth causing further retraction of the tongue, thus reducing the retroglossal dimension. Furthermore, in such patients palatal vibrations could be exaggerated as a result of an open mouth. Improving nasal airflow and congestion could therefore result in more laminar airflow and an improvement in SRBD. Nasal congestion due to any of the above-mentioned factors can interfere with the use of CPAP therapy and reduce compliance or adherence in these patients. Nasal surgery could therefore be instrumental in facilitating CPAP utilisation.

Radiofrequency thermotherapy has been available for nearly 2 decades, and Li et al. [19] first described its use for volumetric reduction of the hypertrophied inferior turbinates. This can often be performed as a clinic-based procedure under local anaesthesia, which could combine topically applied lidocaine followed by injection of 1 or 2% lidocaine. Various appliances are available, including the Somnus®, Celon®, or Coblation® [20]. The aim of the procedure is to create a submucosal thermal lesion, which ultimately results in volumetric reduction and an improved nasal airflow. Great care should be taken to ensure that the mucosa is not breached as this could result in troublesome crusting. Similarly, care should be taken not to insert the probe or the needle too deeply towards the bone of the turbinate as this could potentially lead to necrosis of the bone and possible cacosmia and secondary haemorrhage. A single linear application will generally suffice but 2 applications on each turbinate may be necessary in significantly enlarged turbinates. Some patients may require a further treatment approximately 12 months later. The pain and complications associated with this kind of therapy are relatively mild and the outcome is fairly favourable [20, 21]. A recent study has demonstrated similar outcomes when comparing radiofrequency treatment to bipolar electrocautery application [22].

Nasal surgery alone will rarely remove the requirement for CPAP but may facilitate its use, particularly nasal CPAP. In a recent meta-analysis, the value of nasal surgery was confirmed in reducing CPAP pressure requirements and patient discomfort, although the value of turbinate surgery is less clear [23, 24]. The main limitations with these surgical studies remain their power, level of evidence (typically retrospective level IV), and varying definitions of successful outcomes. Treatment to the inferior turbinate can be conducted on its own or could be combined with septoplasty or radiofrequency treatment of the soft palate and the tongue, or in conjunction with the insertion of a palatal pillar implant as most patients with SRBD exhibit multilevel obstruction [10].

Soft Palate Surgery

Numerous techniques to address the soft palate in patients with SRBD have been described and, broadly speaking, they can be divided into either radical or minimally invasive surgery. The former would include variations of uvulopalatopharyngoplasty (UPPP) and the latter would generally include techniques resulting in no or little change in the overall shape of the soft palate. In essence, the approach to which technique suits the patient is dependent on the individual anatomy, and in general UPPP is more suited if the oropharynx is very crowded with tonsillar hypertrophy, redundant pharyngeal mucosa, and an elongated uvula. DISE reveals different types of collapse, namely anteroposterior, circumferential, or lateral wall, in addition to demonstrating the vigour of palatal vibrations [10]. This may assist in planning the exact procedure required for the individual patient. The severity of SRBD would naturally influence the choice of palatal therapy utilised; for example, patients with moderate or severe OSA may benefit more from radical therapy such as UPPP rather than something minimally invasive. In general, minimally invasive palatal surgery is better suited for simple snorers and perhaps mild OSA patients who do not have a very crowded oropharynx. The types of minimally invasive palatal therapy available and that will be included in this discussion are injection sclerotherapy, palatal pillar implants, and radiofrequency surgery.

Injection Sclerotherapy

The simplest minimally invasive surgical procedure available to address the laxity of the soft palate in simple snorers as opposed to snorers with OSA is injection sclerotherapy or injection snoreplasty. The aim of this technique is essentially to inject chemicals into the soft palate and create stiffening, thereby reducing the palatal vibrations. Historically, the first description of this technique was made in 1943, when fish oil was used to treat snoring [25]. Unlike radiofrequency, this technique can only be used on the soft palate and not on the inferior turbinate or the tongue.

The procedure is relatively simple and is clinic based. Local anaesthesia in the form of benzocaine topical spray or gel is initially used to reduce the gag reflex, followed by injection of 1 or 2% lidocaine. However, some would consider the injection of lidocaine unnecessary as the procedure entails pain created by the insertion of the needle, which is not indifferent. A number of different sclerosing agents have been used for this purpose and include sodium tetradecyl sulphate and aethoxysklerol [26–28]. A single injection of 1.5–2 mL is usually applied submucosally in the midline, approximately 1 cm away from the junction of the hard and soft palate. If this partially resolves the snoring, a further treatment application can be considered a few weeks later, on which occasion the injection is applied in the lateral aspects of the previous midline site. Care should be taken to stay away from the base of the uvula to avoid excessive swelling and also not to apply the injection too deeply. The aim of this treatment is to cause superficial sloughing, which is subsequently replaced by scar tissue that creates the desired stiffening of the soft palate. This form of therapy has been linked with good short-term outcomes in 76–92% of patients and, apart from commonly encountered ulceration and sloughing, complications such as palatal fistula occur in less than 1% of patients [26].

Palatal Implant

In the late 1990s, radiofrequency treatment therapy to the soft palate was introduced and gained interest and acceptance, but it became apparent that this therapy commonly required repeated applications [29]. With this in mind, newer concepts were being developed to combine minimally invasive therapy with a single application to attain stiffening of the soft palate. To this effect, the Pillar® system was introduced. This involved inserting cylindrical implants measuring 18 × 1.5 mm within the substance of the soft palate. These implants are made of a braided segment of a linear polyester called polyethylene terephthalate and have been advocated for their ability to enhance a fibrotic response and for being biostable [30]. The implants are deployed into the soft palate via a disposable delivery kit which has an insertion needle curved appropriately to accommodate the curvature of the soft palate (Fig. 1). The procedure is performed under local anaesthesia after cleaning of the soft palate with chlorhexidine or a similar substance. A topical, injected application of local anaesthesia is used as described earlier. The soft palate should be broad and long enough (at least 25 mm) to accommodate the implants. The insertion point should be as close as possible to the junction of the hard and the soft palate, thus the implants act as an extension of the hard palate and reduce the amount of soft palate that is potentially free to vibrate. The first implant is deployed in the midline followed by 1 on either side, approximately 2 mm apart from each other (Fig. 1). Great care must be taken to ensure that the implant is well positioned within the muscle layer. Preliminary studies were conducted on primary snorers and short-term outcomes in relation to the subjective reduction in snoring reported by partners were encouraging [31–34].

Complications of Pillar implants include extrusion of the implant, which is relatively straightforward to deal with if the extrusion is anteriorly in the oral cavity, but more challenging if it extrudes posteriorly. A recent study confirmed that there are no detrimental effects on the Eustachian tube or middle ear function secondary to altering the rigidity of the palate [35]. In patients with primary snoring, the Pillar implant can be used on its own, but in patients with more severe and intense SRBD, it may be used in conjunction with other treatment modalities [36].

Radiofrequency Surgery

Interstitial radiofrequency application to the soft palate using the Somnus® system was first described by Powell et al. [37]. This system utilised the monopolar principal, but many other devices have subsequently been introduced that utilise bipolar technology, with some acting in wet medium only. These include Coblator®, Celon®, Sutter®, and Ellman®. The setting of energy delivery is different with the individual system used. Some systems allow only interstitial thermotherapy, whereas others offer an electrosection in addition to resecting part of the uvula or redundant web-

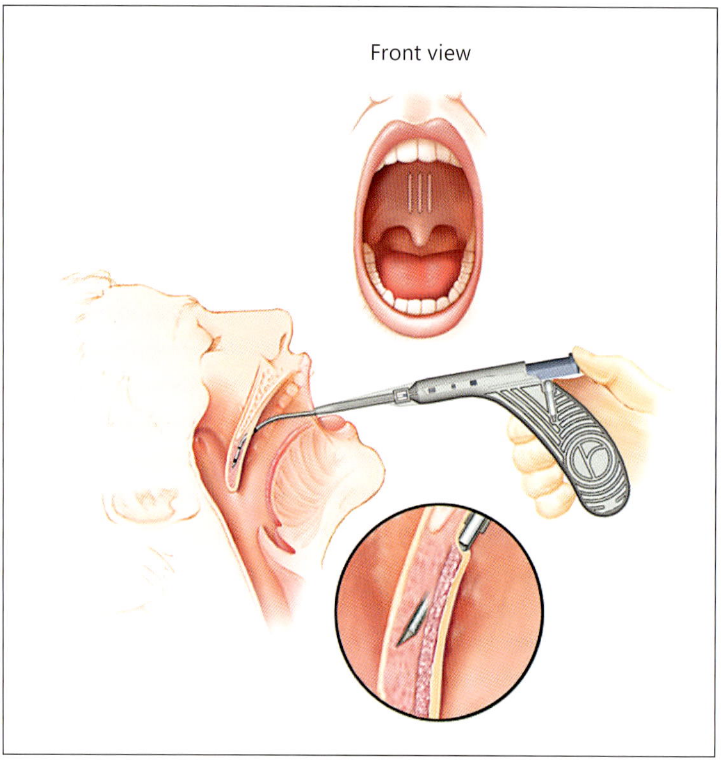

Fig. 1. The Pillar implant delivery system and insertion of implants.

bing at the free edge of the palate. In general, the bipolar systems utilise less power and tend to cause less heat energy to be dissipated and, therefore, thermal energy being delivered more accurately. The advantage of radiofrequency over the laser as a surgical tool at the cellular level (Fig. 2) has been demonstrated by the author [38].

Initial descriptions of probe insertion were fairly conservative, with early reports suggesting 3 applications and then moving on to 6 with bipolar devices, and we recently reported the use of up to 10 applications (Fig. 3) on the soft palate with the Celon® bipolar device [39]. Efficacious outcomes with minimal morbidity have similarly been described with the cold ablation (Coblation) technique, which also offers the advantage of a multifunctional wand applicator that integrates saline infusion, suction attachment, and power controls [40, 41]. Some long-term data are supportive for the use of radiofrequency ablation in mild and moderately severe OSA with significant objective improvements and low morbidity [42].

Thermotherapy to the soft palate can quite easily be conducted under local anaesthesia; however, in many cases general anaesthesia may be preferred as multilevel surgery is the norm in many of these patients with SRBD [39]. The incidence of complications associated with radiofrequency palatoplasty are relatively low and include minor mucosal ulceration, mild velopharyngeal insufficiency, haemorrhage, and, if severe infection sets in or aggressive application of thermotherapy has been applied, a palatal fistula can occur [43, 44]. More often than not, it is necessary to repeat the radiofrequency application to the soft palate at some stage as there may be a recurrence of the symptoms with the elapse of time as there may be some regeneration of the elastic fibres of the previously stiffened soft palate [44].

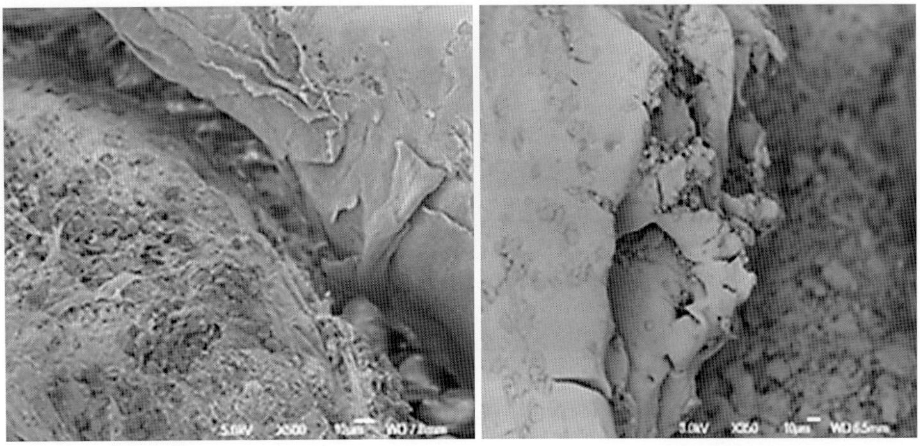

Fig. 2. Scanning electron micrographs demonstrating a highly uniform cellular pattern on the left with radiofrequency excision of the uvula, and a more destructive pattern seen on the right with the use of a laser.

Tongue Base Surgery

Minimally invasive surgery in relation to the hypopharynx mainly involves the base of the tongue and in general is performed to cause a volumetric reduction of the base of the tongue. Again, as in the case of the soft palate, there are various energy delivery systems available and the most commonly utilised are Coblation®, Celon®, and Somnus®. Although the procedure can be performed under local anaesthesia with or without sedation, the author is of the opinion that general anaesthesia would be preferred by patients and offers a better access, therefore attaining a more comprehensive application of thermotherapy [39]. Tongue base surgery alone may not sufficiently resolve the symptoms and quite often is combined with surgery to another anatomical region, and hence the preference of general anaesthesia is justified.

In performing surgery to the base of the tongue, care must be taken to avoid damage to the neurovascular bundles, and hence avoid the lateral (1–1.5 cm) borders of the tongue. Various techniques have been described and include the transcervical approach, ventral and dorsal application through the oral cavity, as well as in combination with the SMILE (intraoral submucosal minimally invasive lingual excision) procedure [39, 45–47].

Ultrasound Doppler probes can be used to identify the lingual artery and avoid trauma to it by keeping more than 5 mm medial to it, and the Coblation plasma wand can be utilised to remove and coagulate the tissue of the tongue as necessary. More simply, the bipolar Celon application or the monopolar Somnus can be utilised in the dorsal aspect of the posterior third of the tongue to create volumetric reduction.

Numerous studies have shown that the short-term success rates with radiofrequency surgery are not dissimilar to those of the more invasive tongue base resection [48, 49]. Complications arising from radiofrequency surgery to the tongue base are rare and include ulceration or abscess formation, mild taste disturbance, temporary tongue numbness, and, rarely, a submental abscess in the region of the thyroglossal cyst [50]. The use of prophylactic antibiotics may reduce the rates of infection and abscess formation.

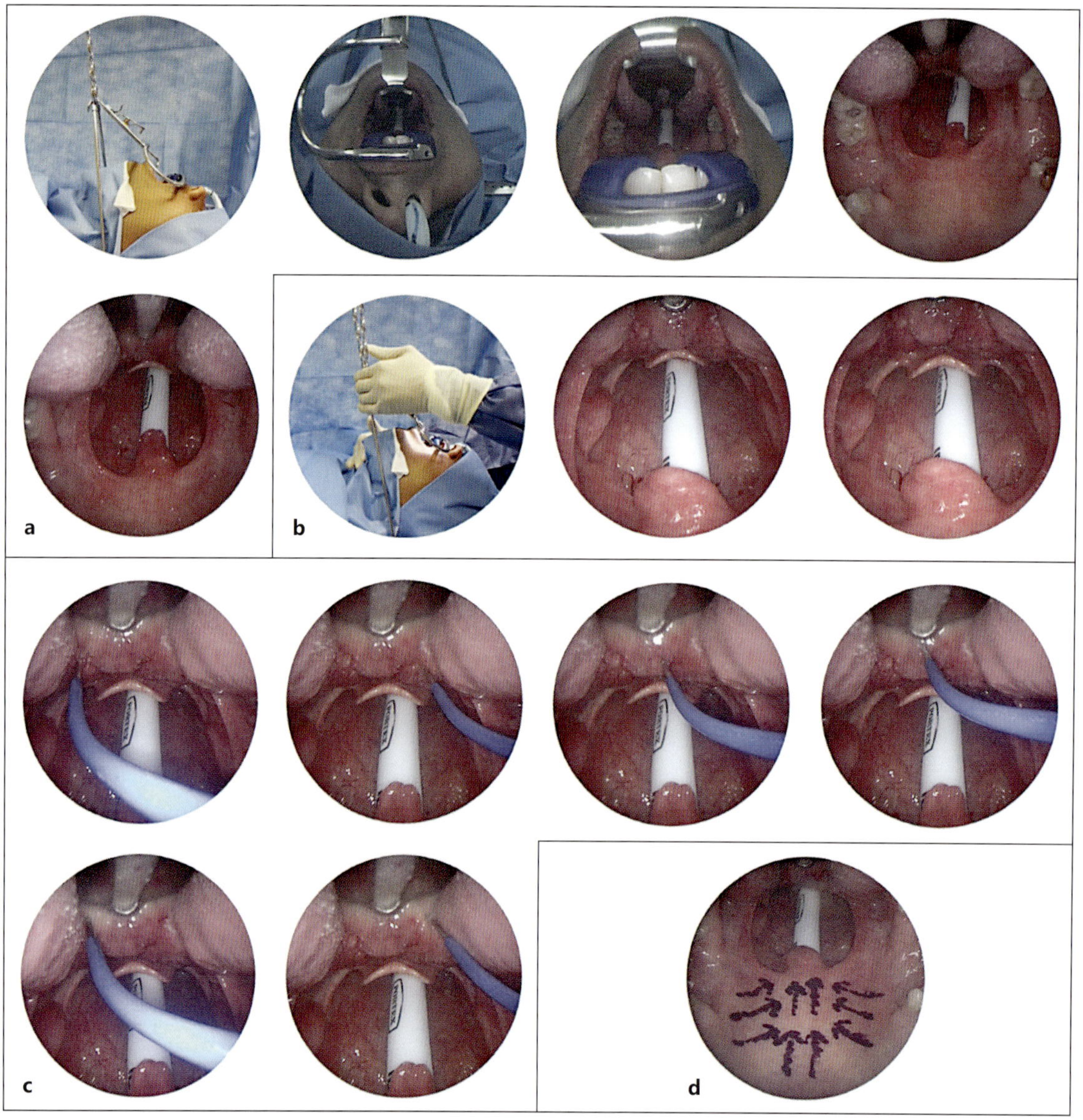

Fig. 3. a The setup for multilevel radiofrequency surgery. **b** Nasotracheal tube with the epiglottis seen clearly. **c** Different probe insertion points for the base of the tongue. **d** Arrows demonstrating radiofrequency probe insertion points for the soft palate.

Conclusion

Minimally invasive surgery for SRBD now forms a frequent and useful adjunct in the management of these patients. Multilevel surgery is often indicated as it is now well documented that many patients with SRBD exhibit anatomical upper airway obstruction at various levels. A multidisciplinary approach to management is advocated and the role of surgical intervention is helpful in patients experiencing difficulty with oral appliances or CPAP therapy.

References

1 Kotecha BT, Hall AC: Role of surgery in adult obstructive sleep apnoea. Sleep Med Rev 2014;18:405–413.
2 Virk JS, Kotecha B: Otorhinolaryngological aspects of sleep-related breathing disorders. J Thoarc Dis 2016;8:9–17.
3 Kakkar RK, Berry RB: Positive airway pressure treatment for obstructive sleep apnoea. Chest 2007;132:1057–1072.
4 National Institute for Health and Care Excellence: Continuous positive airway pressure for obstructive sleep apnoea/hypopnea syndrome. https://www.nice.org.uk/guidance/tal39/resources/continuous-positive-airway-pressure-for-obstructive-sleep-apnoeahypopnea-syndrome-374791501.
5 Georgalas C, Garas G, Hadjihannas E, et al: Assessment of obstruction level and selection of patients for obstructive sleep apnoea surgery: an evidence-based approach. J Laryngol Otol 2010;124:1–9.
6 Abdullah VJ, Lee DL, Ha SC, van Hasselt CA: Sleep endoscopy with midazolam: sedation level evaluation with bispectral analysis. Otolaryngol Head Neck Surg 2013;148:331–337.
7 Babar-Craig H, Rajani NK, Bailey P, Kotecha BT: Validation of sleep nasendoscopy for assessment of snoring with bispectral index monitoring. Eur Arch Otolaryngol 2012,269:1277–1279.
8 Kezirian EJ, White DP, Malhotra A, et al: Inter-rater reliability of drug-induced sleep endoscopy. Arch Otolaryngol Head Neck Surg 2010;136:393–397.
9 Hewitt RJD, Dasgupta A, Singh A, Dutta C, Kotecha B: Is sleep nasendoscopy a valuable adjunct to clinical examination in the evaluation of upper airway obstruction? Eur Arch Otolaryngol 2009; 266:691–697.
10 Kotecha BT, Hannan AS, Khalil HMB, Georgalas C, Bailey P: Sleep nasendoscopy: a 10 year retrospective audit study. Eur Arch Otolaryngol 2007;264:1361–1367.
11 Kamami Y: Laser CO_2 for snoring, preliminary results. Acta Otorhinolaryngol Belg 1990;44:45–56.
12 Ryan CF, Love LL: Unpredictable results of laser assisted uvulopalatoplasty in treatment of obstructive sleep apnoea. Thorax 2000;55:399–404.
13 Walker RP, Gopalsami C: Laser-assisted uvulopalatoplasty: postoperative complications. Laryngoscope 1996;106:834–838.
14 Kotecha B: The nose, snoring and obstructive sleep apnoea. Rhinology 2011; 49:259–263.
15 Young T, Finn L, Palta M: Chronic nasal congestion at night is a risk factor for snoring in a population-based cohort study. Arch Inter Med 2001;161:1514–1519.
16 Virkkula P, Bachour A, Hytonen M, Malamberg H, SalmiT, Maasilta P: Patient and bed partner reported symptoms, smoking and nasal resistance in sleep-disordered breathing. Chest 2005; 128:2176–2182.
17 Kohler M, Clarenbach CF, Thurnheer R, Bloch KE: Perception of nocturnal nasal obstruction in chronic rhinitis. Eur Respir J 2005;26(suppl 49):565.
18 Fergusson BJ: Influence of allergic rhinitis on sleep. Otolaryngol Head Neck Surg 2004;130:617–629.
19 Li KK, Powell NB, Riley RW, Troell RJ, Guilleminault C: Radiofrequency volumetric tissue reduction for treatment of turbinate hypertrophy – a pilot study. Otolaryngol Head Neck Surg 1998;119: 569–573.
20 Bhattacharyya N, Kepnes LJ: Clinical effectiveness of coblation inferior turbinate reduction. Otolaryngol Head Neck Surg 2003;129;365–371.
21 Atef A, Mosleh M, El Bosraty H, et al: Bipolar radiofrequency volumetric tissue reduction of inferior turbinate: does the number of treatment sessions influence the final outcome? Am J Rhinol 2006;20:25–31.
22 Uluyol S, Karakaya NE, Gur MH, Kilicaslan S, Kantarcioglu EO, Yagiz O, Arslan IB: Radiofrequency thermal ablation versus bipolar electrocautery for inferior turbinate hypertrophy: comparison of efficacy and postoperative morbidity. Int Arch Otolaryngol 2016;20: 2–5.
23 Camacho M, Riaz M, Capasso R, et al: The effect of nasal surgery on continuous positive airway pressure device use and therapeutic treatment pressures: a systematic review and meta-analysis. Sleep 2015;38:279–286.
24 Camacho M, Zaghi S, Tran D, Song SA, Chang ET, Certal V: Inferior turbinate size and CPAP titration based treatment pressures: no association found among patients who have not had nasal surgery. Int J Otolaryngol 2016;2016:5951273.
25 Sratus JF: A new approach to the treatment of snoring. Arch Otolaryngol 1943; 38:225–229.
26 Brietzke SE, Mair EA: Injection snoreplasty: extended follow-up and new objective data. Otolaryngol Head Neck Surg 2003;128:605–615.
27 Iseri M, Balcioglu O: Radiofrequency versus injection snoreplasty in simple snoring. Otolaryngol Head Neck Surg 2005;133:224–228.

28 Olszewska E, Panek J, O'Day J, Rogowski M: Usefulness of snoreplasty in the treatment of snoring and mild obstructive sleep apnoea/hypopnea syndrome – preliminary report. Otolaryngol Pol 2014;68:184–188.
29 Li KK, Powell NB, Riley RW, et al: Radiofrequency volumetric reduction of the palate: an extended follow-up study. Otolaryngol Head Neck Surg 2000;122: 410–414.
30 Greenwald D, Shunway S, Albear P, et al: Mechanical comparison of 10 suture materials before and after in vivo incubation. J Surg Res 1994;56:372–377.
31 Malhotra A, Pillar G, Fogel R, et al: Upper airway collapsibility: measurement and sleep effects. Chest 2001;120:156–161.
32 Ho WK, Wei WI, Chung KF: Managing disturbing snoring with palatal implants: a pilot study. Arch Otolaryngol Head Neck Surg 2004;130:753–758.
33 Friedman M, Vidyasagar R, Bliznikas D, Joseph NJ: Patient selection and efficacy of Pillar implant technique for treatment of snoring and obstructive sleep apnea/hypopnea syndrome. Otolaryngol Head Neck Surg 2006;134:187–196.
34 Maurer JT, Verse T, Stuck BA, Hormann K, Hein G: Palatal implants for primary snoring: short-term results of a new minimally invasive surgical technique. Otolaryngol Head Neck Surg 2005;132: 125–131.
35 Lin HC, Friedman M, Chang HW, Lai CC, Chin CH, Lin MC, Pott T, Samuelson C: Effects of Pillar implants for sleep-related breathing disorders on middle ear function. Eur Arch Otolaryngol 2013;270:2339–2343.
36 Zhang XM, Tham CJ, Yin YL, Sun YQ, Zhou X: A novel palatal implant surgery combined with uvulopalatopharyngoplasty and inferior turbinate radiofrequency for the treatment of moderate to severe obstructive sleep apnoea: a pilot study. Eur Arch Otorhinolaryngol 2015; 272:1195–1202.
37 Powell NB, Riley RW, Troell RJ, Li K, Blumen MB, Guilleminault C: Radiofrequency volumetric tissue reduction of the palate in subjects with sleep-disordered breathing. Chest 1998;113:1163–1174.
38 Stimpson P, Kotecha B: Histopathological and ultrastructural effects of cutting radiofrequency energy on palatal soft tissues: a prospective study. Eur Arch Otolaryngol 2011;268:1829–1836.
39 Virk JS, Nouraei R, Kotecha B: Multilevel radiofrequency ablation to the soft palate and tongue base: tips and pitfalls. Eur Arch Otolaryngol 2014;271:1809–1813.
40 Tvinnereim M, Mitic S, Hansen RK: Plasma radiofrequency preceded by pressure recording enhances success for treating sleep-related breathing disorders. Laryngoscope 2007;117;731–736.
41 Milkov M, Nedev P, Tonchev T, Kirov F, Madjova H: Usage of Coblator-2 (Arthrocare) and ENT Celon (Olympus) systems for soft palate reduction in habitual snoring patients. Sleep Med 2013; 14:e209.
42 Veer V, Yang W-Y, Green R, Kotecha B: Long-term safety and efficacy of radiofrequency ablation in the treatment of sleep disordered breathing: a meta-analysis. Eur Arch Otolaryngol 2014;271: 2863–2870.
43 Kezirian EJ, Powell NB, Riley RW, Hester JE: Incidence of complications in radiofrequency treatment of the upper airway. Laryngoscope 2005;115:1298–1304.
44 Virk JS, Kumar G, Al-Okati D, Kotecha B: Radiofrequency ablation in snoring surgery: local tissue effects and safety measures. Eur Arch Otolaryngol 2014; 271:3313–3318.
45 Robinson S, Lewis R, Norton A, McPeake S: Ultrasound-guided radiofrequency submucosal tongue-base excision for sleep apnoea: a preliminary report. Clin Otolaryngol Allied Sci 2003; 28:341–345.
46 Maturo SC, Mair EA: Submucosal minimally invasive lingual excision (SMILE): an effective, novel surgery for pediatric tongue base reduction. Ann Oto Rhino Laryngol 2006;115:624–630.
47 Robinson S, Ettema SL, Brusky L, Woodson BT: Lingual tonsillectomy using bipolar radiofrequency plasma excision. Otolaryngol Head Neck Surg 2006;134:328–330.
48 Stuck BA, Maurer JT, Verse T, Hormann K: Tongue base reduction with temperature-controlled radiofrequency volumetric tissue reduction for treatment of obstructive sleep apnoea syndrome. Acta Otolaryngol 2002;122:531–536.
49 Li KK, Powell NB, Riley RW, Guilleminault C: Temperature-controlled radiofrequency tongue base reduction for sleep-disordered breathing: long-term outcomes. Otolaryngol Head Neck Surg 2002;127:230–234.
50 Tornari C, Wong G, Arora A, Kotecha B: A unique complication of radiofrequency therapy to the tongue base. Int J Surg Case Rep 2015;8C:9–12.

Prof. Bhik Kotecha, MBBCH, MPhil, FRCS, DLO
Royal National Throat, Nose and Ear Hospital (UCLH)
330 Grays Inn Road
London WC1X 8DA (UK)
E-Mail bhikkot@aol.com

Lin H-C (ed): Sleep-Related Breathing Disorders. Adv Otorhinolaryngol. Basel, Karger, 2017, vol 80, pp 99–108
DOI: 10.1159/000470873

Advances in Box Surgery for Obstructive Sleep Apnea: Genioglossus Advancement, Hyoid Suspension, and Maxillomandibular Advancement

Yau Hong Goh[a] • Winston Tan[b] • Victor James Abdullah[c] • Sung Wan Kim[d]

[a]Ear Nose Throat, Head and Neck Surgery and [b]Oral Maxillofacial Surgery, Mount Elizabeth Medical Centre, Singapore, Singapore; [c]Department of Otorhinolaryngology, Head and Neck Surgery, United Christian Hospital, Kowloon East Cluster, Hong Kong, SAR, China; [d]Department of Otorhinolaryngology-Head and Neck Surgery, School of Medicine, Kyung Hee University, Seoul, Republic of Korea

Abstract

Box surgery for obstructive sleep apnea (OSA) patients consists of mandibular surgery, including genioglossus advancement, hyoid suspension, and maxillomandibular advancement, as an airway reconstruction. This surgery was developed in the early 1980s. Thereafter, techniques were modified in each surgery for the enhancement of outcome and prevention of complications. However, the indication for surgery remains poorly defined due to the dynamicity of the upper airway, variability of the phenotype in OSA patients, and absence of a representative method for the obstruction site in the upper airway. As a result, box surgery is performed restrictively. On the development of an evaluation method including radiologic and endoscopic examination during sleep, the indications and surgical outcomes of each box surgery should be standardized and reevaluated. In this review, the development and limitations of box surgery will be discussed for the positive future of this surgery.

Genioglossus Advancement

The upper airway muscles are responsible for the maintenance of the upper airway during sleep. Among them, the genioglossus muscle (GGM) is the major dilator muscle. The GGM has tonic and reflexive activation during sleep and wake [1]. Its activity is greater in obstructive sleep apnea (OSA) than in normal controls, which may lead to adaptive changes in the distribution of muscle fiber types. These findings may be suggested as a rationale for the advancement and reposition of the GGM [2].

In preoperative evaluations, locating the genial tubercle before surgery is mandatory and was initially assessed with a simple radiograph, such as a cephalogram and panoramic radiograph. However, obtaining the precise location of the genial tubercle is not sufficient, especially when there has been no mention of the definite anatomical advantage of locating it during the surgery. According to an analysis of CT scans, it was recently reported that the position of the genial tubercle is highly variable. Thus, a 3-dimensional CT scan analysis may help to increase the likelihood of locating the tubercle accurately. According to the analysis, almost 30% of the population may be contraindicated for genioglossus advancement (GGA) due to the short distance between the dental root and genial tubercle, which results in possible dental injury during osteotomy [3]. Dental caries or apical abscesses should also be evaluated preoperatively because they can be a source of infection.

The surgical technique of GGA involves performing a mandibular osteotomy with anterior repositioning of the segment of the mandible attached to the genioglossus. In 1983, Powell et al. [4] introduced an intraoral horizontal sliding osteotomy of the anterior mandible to advance the genioglossus, mylohyoid, and anterior digastric muscles along with bilateral sagittal split osteotomy. Three years later, they reported an inferior sagittal osteotomy which was a modified version of their original procedure. However, it was abandoned due to the risk of mandibular fracture [5]. In 1990, they reported a modified method – geniotubercle advancement (GTA). A rectangular osteotomy of the anterior mandible, including the genial tubercle, was performed, then advanced and rotated 60 to 90° prior to fixation [6]. This technique is still widely used (Fig. 1), and the procedure has gradually advanced with various modifications.

Among the modifications suggested, the trephine osteotomy approach has been introduced. A circular osteotomy segment containing the GGM in 1 cutting motion was obtained using a commercially available surgical set. The method managed to shorten the average operating time and also reduce the amount of total blood loss. The surgical result as a part of multilevel surgery was also almost similar to that of GTA [7]. A mortised genioplasty, which allows massive movement of the GGM and suprahyoid muscles, has also been introduced. This technique may be the most powerful method to advance the GGM together with the suprahyoid muscle, and thus an optimum outcome was anticipated. However, this technique presents the limitation of weakening the mandible, which makes postoperative mandible fracture a possibility. Too much tension to the advanced GGM may create muscle fatigue and result in relapse [8]. To prevent a weak mandible, elliptical window GGA was reported for stabilizing the mandible [9]. To avoid rotation of the GGM during advancement, several modifications were reported in succession. In some of these, the advanced geniotomy segment was fixed on the anterior surface of the mandible using a titanium plate or mesh without rotation of the GGM [10]. Trapezoid mortised genioplasty increased the width of the lingual cortex of the geniotomy bony fragment and fixed at the anterior surface of the mandible without rotation of the muscle [11]. Besides these modifications, many surgical approaches have been reported.

Modifications in surgical techniques have been focused on increasing the amount of GGM advancement, avoiding rotation of the GGM, and prevention of major complications, such as mandible fracture. However, all GGA procedures may have possible sequelae, including numbness of the skin on the mandible, a change in skin color, and distortion of dentition.

The surgical outcome of GGA is variable. Success rates are reported to be between 39 and 78% [12], and are heavily influenced by multilevel surgery rather than GGA alone. These variable results may be due to different surgical indications, different surgical techniques, and possibly an incorrect surgical approach secondary to inappropriate evaluation. GGA has been advocated by many physicians for OSA patients with tongue base obstruction. However, the indications for performing GGA in these patients, which involves the type and level of tongue base obstruction, are not yet fully developed. Furthermore, GGA is usually part of a multilevel surgery, which

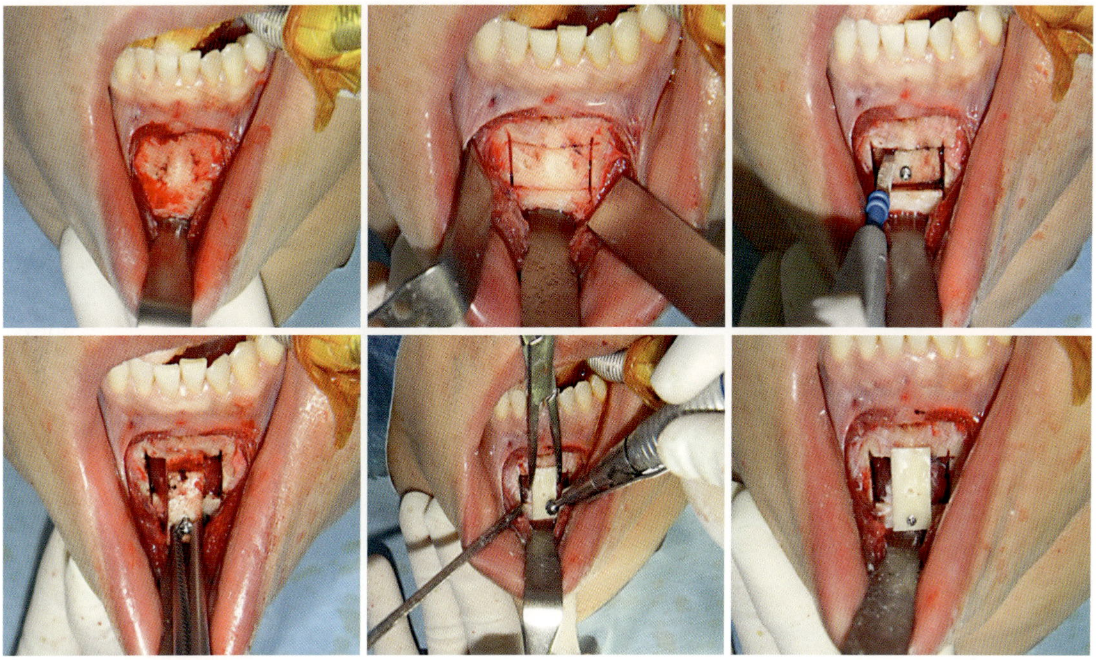

Fig. 1. Geniotubercle advancement.

means the benefit from GGA alone is so far not well established. The urgent issues involved in GGA are to define the most suitable type of tongue base obstruction, and criteria for accurate patient assessment and evaluation so that GGA is performed correctly, and to obtain a surgical outcome reflecting the effect of GGA alone.

Hyoid Suspension

The hyoid cartilage ossifies in 6 centers: 2 in the body, and 1 in each cornu during fetal development and the first 2 years of life. The hyoid bone provides attachments to muscles of the tongue (hyoglossus, genioglossus, and the intrinsic muscles of the tongue), the suprahyoid muscles (digastric muscle, stylohyoid muscle, geniohyoid muscle, and the mylohyoid muscle), and inferiorly, the strap muscles (thyrohyoid muscle, omohyoid muscle, and the sternohyoid muscle). The greater cornua also provides attachments to the middle constrictor muscle (Fig. 2).

It is understandable that the hyoid bone has been of much interest in the surgical treatment of OSA in order to expand the pharynx, in particular the hypopharynx and the retrolingual airway when obstructions at these sites are suspected. It is evident that the hypopharyngeal airway can decrease by almost a third with simple postural changes, when the patient moves from the upright to the supine position [13]. In canine studies in the early 1980s, using a stainless steel brace, it was demonstrated that anterior positioning of the hyoid body and the lateral displacement of the greater cornua can expand the hypopharyngeal and tongue base airway [14].

The significance of hyoid suspension, probably via its effect on the hyoglossus muscle and the intrinsic and middle constrictor muscles, was well

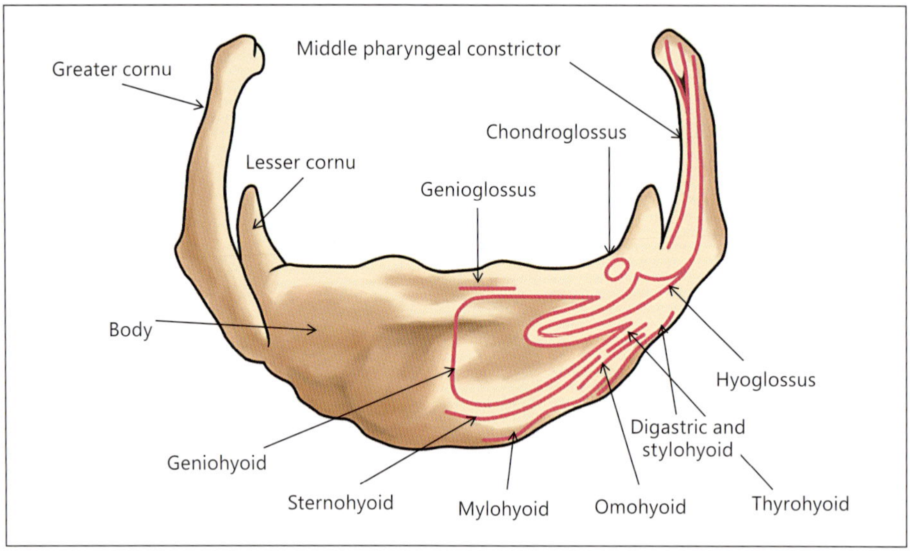

Fig. 2. The hyoid bone with illustrated muscle attachments.

appreciated by Riley and Powell [15, 16] in the early 1980s. Their preferred technique was hyoid suspension to the lower border of the mandible using the fascia lata with the division of the infrahyoid muscles. The procedure was usually performed as a part of multilevel airway expansion surgery via mandibular or maxillomandibular osteotomies and their advancement. In the early 1990s, Riley and Powell modified their technique to hyoid-thyroid-pexia, which will here be referred to as hyothyroidopexy. The procedure involves the suturing of the hyoid bone to the thyroid cartilage with the partial release of the suprahyoid muscles and the division of the stylohyoid ligaments attached to the lesser cornua [17] (Fig. 3).

The author's (V.J.A.) preferred technique is hyothyroidopexy via a transverse skin crease incision placed halfway between the hyoid bone and the thyroid notch. The platysmal flaps are raised both superiorly and inferiorly, which can be retracted with stay sutures. The hyoid bone is clearly demarcated with monopolar diathermy via the release of tissues and limited release of the suprahyoid muscle fibers. The stylohyoid muscles at our institution are not routinely divided. The superior border of the thyroid cartilage is clearly identified by dividing the sternohyoid strap muscle fibers. Our preferred sutures are 2 Ethibond V37 polyester sutures on a 40-mm tapercut needle. Four sutures are applied through the upper thyroid cartilage, with 2 on each thyroid ala. Each needle is passed deep into the hyoid and by hugging its deep surface the tip is gently rotated through, just above the hyoid. All 4 sutures are passed before tying. The apposition of the hyoid and thyroid cartilage is facilitated by the surgical assistant with a pair of Littlewood tissue forceps on the body of the hyoid, displacing it caudally while gently pushing the larynx upwards with his or her other hand. The 4 sutures can then be tied snuggly (Fig. 4). Complications are rare and some temporary discomfort in swallowing and feeling of neck tightness may be experienced by the patient, which either settles or the patient adapts to with time [18].

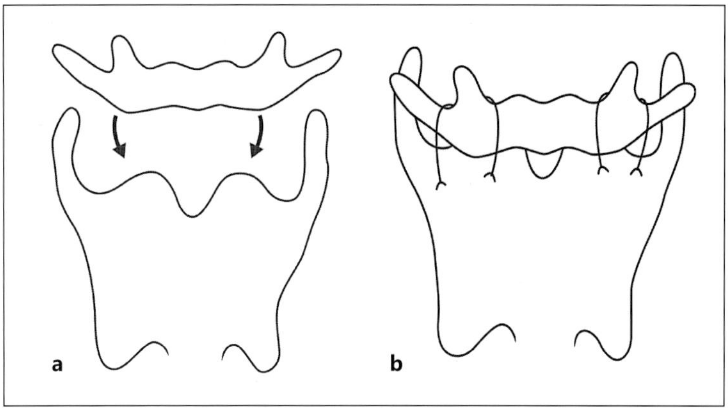

Fig. 3. Hyothyroidopexy: before surgery (**a**) and after surgery (**b**).

There are alternative hyothyroidopexy techniques. Hormann and Baisch [19] utilize a stainless steel wire that sometimes cuts through the cartilage. This technique was modified in Italy by Piccin et al. [20] using 2-mm titanium plates on the thyroid cartilage and passing the wire through the plate instead. Hyoid expansion using an implantable device (the Air Frame system) has been examined in an attempt to open up the hypopharynx, but this did not result in a significant improvement in sleep parameters [21]. The above-described technique remains the author's preference.

To date, it is difficult to ascertain the exact contribution of the hyoid suspension or hyothyroidopexy in a multilevel surgical procedure for OSA as prospective randomized studies are understandably lacking in OSA patients. A systematic review and meta-analysis by Song et al. [22] was helpful in putting the available information into perspective. A total of 498 studies were screened and 64 were reviewed, of which 9 met the criteria for inclusion with a total of 101 patients. In patients undergoing isolated hyoid surgery for OSA, hyoid suspension was able to reduce the apnea-hypopnea index (AHI) by 38.3%, whereas for hyothyroidopexy an AHI reduction of 50.7% was achieved. As for isolated hyoid expansion, a 7.1% improvement in AHI was observed. This information indicates that hyothyroidopexy is likely to contribute to the overall improvement as a single procedure or if incorporated in a multilevel procedure, and is probably superior to hyoid suspension despite the hyoid usually being situated at a lower level in many OSA patients [23]. The systematic review and meta-analysis is of course limited by the quality of the published studies available, most of which were case control or retrospective case series. In 2005, den Herder et al. [24] revealed that, in their series, hyothyroidopexy alone as a primary procedure achieved better results than when it was done as a secondary procedure, for example to salvage a uvulopalatopharyngoplasty failure. The residual obstruction in the secondary cases was likely to have altered the laxity of tissues and the degree of tissue swelling, thus limiting the response. Isolated primary hyoid surgery is seldom performed for OSA patients and the effect of the likely preferred hyothyroidopexy in a multilevel procedure requires further evaluation.

Maxillomandibular Advancement

Orthognathic surgery was developed in the 1960s for the treatment of dentofacial deformities and dental malocclusion. Its application was later ex-

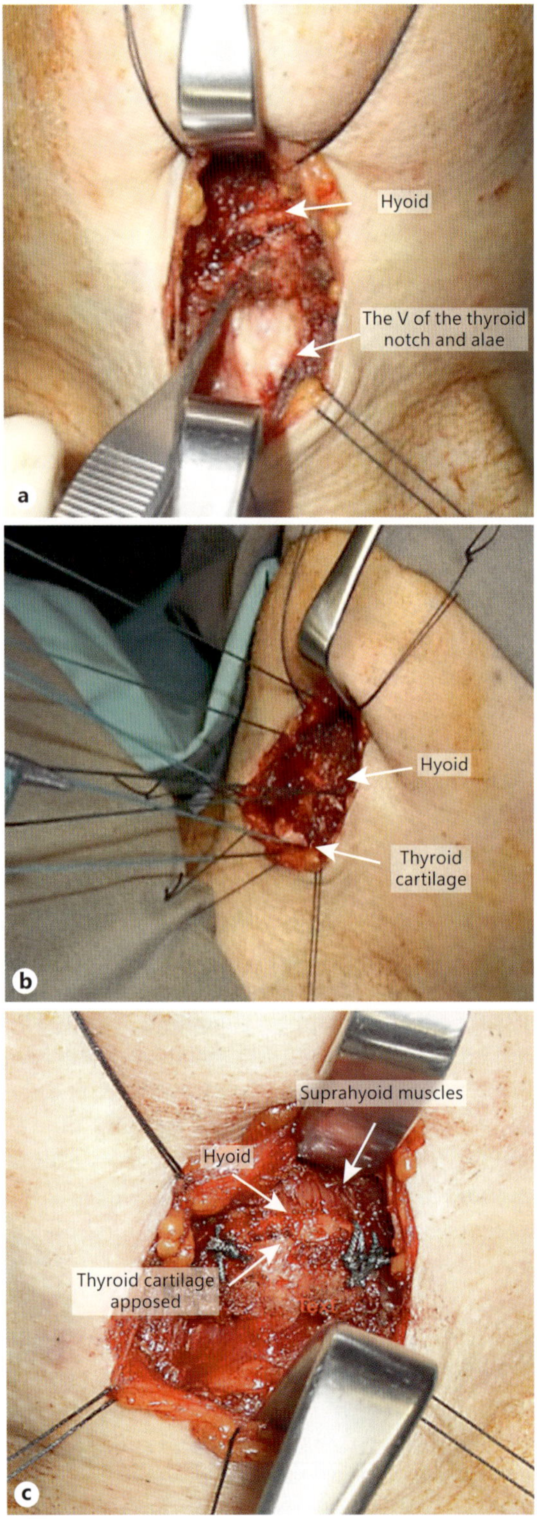

panded to treat patients with severe OSA in the mid 1980s [25, 26]. In the classical phase II surgery for OSA described by Riley and Powell, a 10-mm advancement of the maxilla and mandible resulted in an impressive 97% cure rate in patients who had failed phase I surgery, and 91% in patients treated solely by phase II surgery [27]. However, the results of maxillomandibular advancement (MMA) techniques from various centers around the world have been varied. Reports of varying success rates ranging from 65 to 100% may be attributed mainly to the difference in the definition of success [25–31].

Due to its radical nature, MMA is frequently relegated as a last resort option for the treatment of OSA. While patients with mild degrees of OSA may be treated with soft tissue or limited bone surgeries, those with significant OSA are unlikely to benefit from such interventions alone. For patients with maxillofacial skeletal deficiencies or gross skeletal/soft tissue disproportions, MMA should be considered as a first-line treatment of choice.

MMA is essentially a multilevel surgery. Its efficacy lies in the simultaneous expansion of the retropalatal and retrolingual space, and tightening of the lateral pharyngeal laxity [32]. A recent large meta-analysis of MMA revealed a surgical success of 85.5% (defined as the percentage of patients with a more than 50% reduction in AHI to below 20) and cure rate of 38.5% (defined as a post-MMA AHI below 5) [33]. Presented with such an overwhelming evidence of the efficacy of MMA, why is it that this technique is still not widely advocated or generally accepted? The first

Fig. 4. Surgical techniques of hyothyroidopexy. **a** Exposure via a horizontal cervical skin crease incision placed midway between the hyoid and thyroid notch. **b** Four stitches, 2 Ethibond, are passed through the upper edges of the thyroid cartilage and then over the body and greater cornua of the hyoid. All 4 sutures are passed prior to tying. **c** The stitches are tied with apposition of the hyoid and thyroid cartilage.

reason is that the treatment options available to any patient are, unfortunately, often limited by the expertise of the medical practitioners themselves. Without cross-disciplinary training and collaboration, many patients frequently end up partially treated with occasional nasal continuous positive airway pressure usage (poor compliance) or limited soft tissue surgeries. The second, and perhaps the biggest challenge, remains the acceptance of potential facial profile alterations after MMA surgery.

Of all the ethnic differences in the cephalometric parameters between Far East Asian and Caucasian subjects, the greater incidence of bimaxillary protrusion in the Far East Asian face is perhaps the most important [34–36]. Considering the relatively greater maxillary and mandibular protrusion in the Asian population and the biased perception by the native Chinese in favor of bimaxillary retrusion, the esthetic effects of an increased bimaxillary face after MMA poses a great challenge for Asian patients [37]. Furthermore, patients with bimaxillary protrusion may have lip seal problems due to the protrusion, as evidenced by the upward strain of the mentalis and increased activity of perioral muscles. The increased tension on advanced maxillary and mandibular segments in a patient with bimaxillary protrusion may also affect the stability of the advancement, which may potentially increase regression and postoperative failure rates. All these factors and the potential long-term stability of the advanced segments due to decreased tension after an anterior segmental setback were considered in the development of the modified technique for the Far East Asian face.

When performing MMA, segmental surgery is generally indicated for patients with bimaxillary protrusion. This involves the posterior setback of the anterior maxilla (premaxilla) and the anterior alveolar segment of the mandible, usually after the extraction of a bicuspid from each quadrant of the dental arch (Fig. 5). Together with the standard LeFort I and bilateral sagittal split osteotomies, the surgery, when applied to patients with OSA, achieves the aim of increasing the posterior airway space without significant aesthetic facial alterations [38]. A preoperative surgery model is routinely used to assess patient suitability. These patients can be identified on physical examination by the presence of lip closure problems, excessive proclination, a wide nasal alar base, excessive show of teeth or gums when smiling, and protrusive lips on facial profile examination and cephalometric measurements.

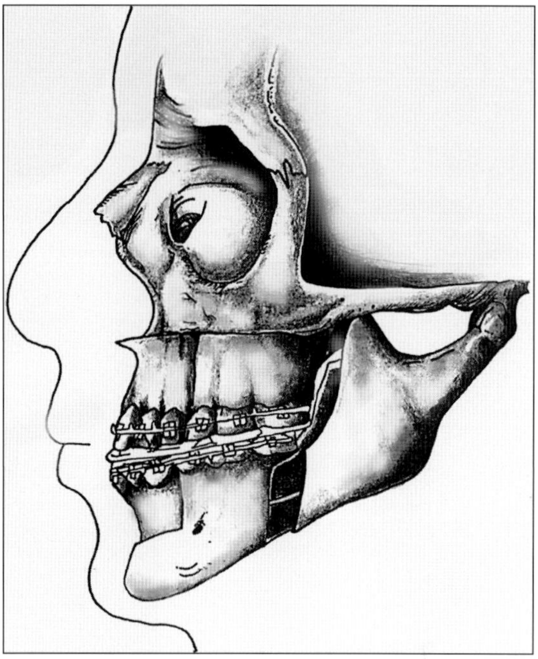

Fig. 5. A diagrammatic depiction of the entire modified maxillomandibular advancement technique: a LeFort I osteotomy with anterior subapical segmental osteotomy and bilateral sagittal split osteotomies.

Our center has been undertaking the modified technique primarily for patients requiring MMA for the treatment of OSA since 2000. The preliminary 4-year review of our first cohort of patients showed encouraging results [39]. The preoperative polysomnography results (±SD) were as follows: apnea index, 55.7 ± 19.8, AHI 70.7 ± 15.9,

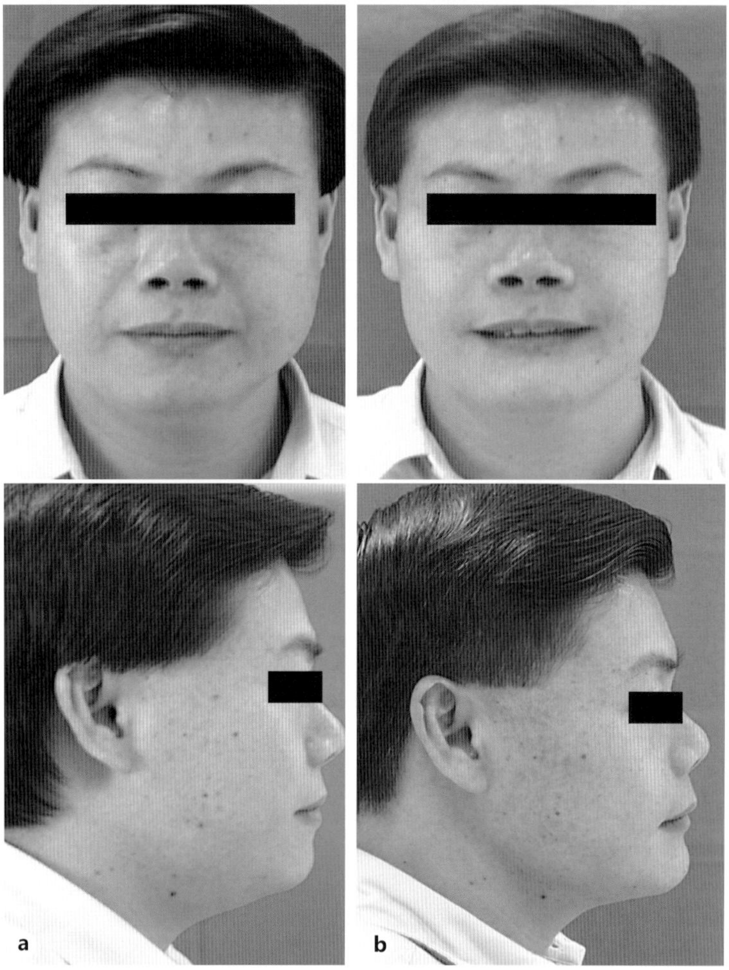

Fig. 6. Facial profile before (**a**) and after (**b**) a modified maxillomandibular advancement technique. This patient presents a good postoperative result, with the apnea-hypopnea index from 53.7 to 0.6 and lowest O_2 saturation from 81 to 92% without a change of facial profile and weight reduction.

lowest oxygen saturation 58.6 ± 12.3%. The postoperative polysomnography results were as follows: apnea index 3.4 ± 3.4, AHI 11.4 ± 7.4, lowest oxygen saturation, 83.9 ± 8.8%. Apart from 2 patients with minor postoperative wound pain that warranted the removal of limited titanium screws and a plate, there were no significant postoperative complications. None of the modified MMA patients experienced any significant facial profile changes after surgery (Fig. 6).

The sustainability of a good treatment outcome, whether medical or surgical, remains the greatest challenge for the clinician. Long-term outcomes are unfortunately rarely reported, even when it is evident in clinical practice that the result of a particular technique deteriorates with time. The importance of a protracted review and follow-up of our posttreatment OSA patients cannot be overemphasized. Over time, the dynamic and fluid nature a patient's anatomy and changes in their lifestyle can greatly affect their initial treatment results.

The ongoing surveillance of our postoperative patients, many of whom were assessed about 10 years after surgery, shows deterioration in their polysomnography parameters. Our currently unpublished data of the first cohort of modified MMA patients showed a clear association between postsurgical weight gain and the negative changes in their polysomnography. Thus, postoperative weight management remains an essential adjunct treatment for the maintenance of a successful outcome.

One of the advantages of the modified MMA technique is the minimal facial profile changes in posttreatment patients. Conceived and designed mainly for the Asian face, the modified MMA technique may be applied to patients from other races. This technique has been shown to be highly effective in treating patients with severe OSA without the significant facial profile alteration commonly associated with the classical MMA. To maintain a long-term outcome, a strict postoperative weight-management program is imperative.

References

1 Horner RL, Hughes SW, Malhotra A: State-dependent and reflex drives to the upper airway: basic physiology with clinical implications. J Appl Physiol 2014;116:325–336.

2 Blumen MB, de La Sota AP, Quera-Salva MA, Frachet B, Chabolle F, Lofaso F: Tongue mechanical characteristics and genioglossus muscle EMG in obstructive sleep apnoea patients. Respir Physiol Neurobiol 2004;140:155–164.

3 Jung SY, Shin SY, Lee KH, Eun YG, Lee YC, Kim SW: Analysis of mandibular structure using 3D facial computed tomography. Otolaryngol Head Neck Surg 2014;151:760–764.

4 Powell N, Guilleminault C, Riley R, Smith L: Mandibular advancement and obstructive sleep apnea syndrome. Bull Eur Physiopathol Respir 1983;19:607–610.

5 Riley RW, Powell NB, Guilleminault C: Inferior sagittal osteotomy of the mandible with hyoid myotomy-suspension: a new procedure for obstructive sleep apnea. Otolaryngol Head Neck Surg 1986;94:589–593.

6 Riley RW, Powell NB, Guilleminault C: Obstructive sleep apnea syndrome: a review of 306 consecutively treated surgical patients. Otolaryngol Head Neck Surg 1993;108:117–125.

7 Miller FR, Watson D, Boseley M: The role of the Genial Bone Advancement Trephine system in conjunction with uvulopalatopharyngoplasty in the multilevel management of obstructive sleep apnea. Otolaryngol Head Neck Surg 2004;130:73–79.

8 Hendler BH, Costello BJ, Silverstein K, Yen D, Goldberg A: A protocol for uvulopalatopharyngoplasty, mortised genioplasty, and maxillomandibular advancement in patients with obstructive sleep apnea: an analysis of 40 cases. J Oral Maxillofac Surg 2001;59:892–897.

9 Dattilo DJ, Aynechi M: Modification of the anterior mandibular osteotomy for genioglossus advancement with hyoid suspension for obstructivesleep apnea. J Oral Maxillofac Surg 2007;65:1876–1879.

10 Mintz SM, Ettinger AC, Geist JR, Geist RY: A modified geniotomy technique for obstructive sleep apnea syndrome. J Oral Maxillofac Surg 1995;53:1226–1228.

11 Singhal D, Hsu SS, Lin CH, Chen YC, Chen YR: Trapezoid mortised genioplasty: a further refinement of mortised genioplasty. Laryngoscope 2013;123:2578–2582.

12 Kezirian EJ, Goldberg AN: Hypopharyngeal surgery in obstructive sleep apnea: an evidence-based medicine review. Arch Otolaryngol Head Neck Surg 2006;132:206–213.

13 Camacho M, Capasso R, Schendel S: Airway changes in obstructive sleep apnoea patients associated with a supine versus an upright position examined using cone beam computed tomography. J Laryngol Otol 2014;128:824–830.

14 Patton TJ, Thawley SE, Waters RC, Vandermeer PJ, Ogura JH: Expansion hyoidplasty: a potential surgical procedure designed for selected patients with obstructive sleep apnea syndrome: experimental canine results. Laryngoscope 1983;93:1387–1396.

15 Riley R, Guilleminault C, Powell N, Derman S: Mandibular osteotomy and hyoid bone advancement for obstructive sleep apnea: a case report. Sleep 1984;7:79–82.

16 Riley RW, Powell NB, Guilleminault C: Inferior sagittal osteotomy of the mandible with hyoid myotomy-suspension: a new procedure for obstructive sleep apnea. Otolaryngol Head Neck Surg 1986;94:589–593.

17 Riley RW, Powell NB, Guilleminault C: Obstructive sleep apnea syndrome: a surgical protocol for dynamic upper airway reconstruction. J Oral Maxillofac Surg 1993;51:742–749.

18 Richard W, Timmer F, vanTinteren H, de Vires N: Complications of hyoid suspension in the treatment of obstructive sleep apnea syndrome. Eur Arch Otorhinolaryngol 2011;268:631–635.

19 Hormann K, Baisch A: How I do it: the hyoid suspension. Laryngoscope 2004;114:1677–1679.

20 Piccin O, Scaramuzzino G, Martone C, Marra F, Gobbi R, Sorrenti G: Modified hyoid suspension technique in the treatment of multilevel related obstructive sleep apnea. Otolaryngol Head Neck Surg 2014;150:321–324.

21 Hamans E, Stuck BA, de Vires N, Boudewyns A, Devolder A, Verbraecken J, van de Heyning P: Hyoid expansion as a treatment for obstructive sleep apnea: a pilot study. Sleep Breath 2013;17:195–201.

22 Song SA, Wei JM, Buttram J, Tolisano AM, Chang ET, Liu SY-C, Certal V, Camacho M: Hyoid surgery alone for obstructive sleep apnea: a systematic review and meta-analysis. Laryngoscope 2016;126:1702–1708.

23 Guilleminault C, Riley R, Powell N: Obstructive sleep apnea and abnormal cephalometric measurements, implications for treatment. Chest 1984;86:793–794.
24 den Herder C, van Tinteren H, de Vries N: Hyoidthyroidpexia: a surgical treatment for sleep apnea syndrome. Laryngoscope 2005;115:740–745.
25 Riley RW, Powell NB, Guilleminault C, Nino-Murcia G: Maxillary-mandibular and hyoid advancement: an alternative to tracheostomy on obstructive sleep apnea. Otolaryngol Head Neck Surg 1986;94:584–588.
26 Waite PD, Wooten V, Lachner J, Gubette RF: Maxillomandibular advancement surgery in 23 patients with obstructive sleep apnea syndrome. J Oral Maxillofac Surg 1989;47:1256–1261.
27 Li KK, Riley RW, Powell NB, Troell R, Guilleminault C: Overview of phase II surgery for obstructive sleep apnea syndrome. Ear Nose Throat J 1999;78:851–857.
28 Hochban W, Conradt R, Brandenburg U, Heitmann J, Peter JH: Surgical maxillofacial treatment of obstructive sleep apnea. Plast Reconstr Surg 1997;99:619–626.
29 Bettega G, Pepin JL, Veale D, Deschaux C, Raphaël B, Lévy P: Obstructive sleep apnea syndrome: fifty-one consecutive patients treated by maxillofacial surgery. Am J Respir Crit Care Med 2000;162:641–649.
30 Prinsell JR: Maxillomandibular advancement surgery in a site-specific treatment approach for obstructive sleep apnea in 50 consecutive patients. Chest 1999;116:1519–1529.
31 Li KK, Riley RW, Powell NB, Guilleminault C: Maxillomandibular advancement for persistent obstructive sleep apnea after phase I surgery in patients without maxillomandibular deficiency. Laryngoscope 2000;110:1684–1688.
32 Liu SY, Huon LK, Iwasaki T, Yoon A, Riley R, Powell N, Torre C, Capasso R: Efficacy of maxillomandibular advancement examined with drug-induced sleep endoscopy and computational fluid dynamics airflow modeling. Otolaryngol Head Neck Surg 2016;154:189–195.
33 Zaghi S, Holty JE, Certal V, Abdullatif J, Guilleminault C, Powell NB, Riley RW, Camacho M: Maxillomandibular advancement for treatment of obstructive sleep apnea: a meta-analysis. JAMA Otolaryngol Head Neck Surg 2016;142:58–66.
34 Gu Y, McNamara JA Jr, Sigler LM, Baccetti T: Comparison of craniofacial characteristics of typical Chinese and Caucasian young adults. Eur J Orthod 2011;33:205–211.
35 Lee JJ, Ramirez SG, Will MJ: Gender and racial differences in cephalometric analysis. Otolaryngol Head Neck Surg 1997;117:326–329.
36 Will MJ, Ester, MS, Ramirez SG, Tiner BD, McAnear JT, Epstein L: Comparison of cephalometric analysis with ethnicity in obstructive sleep apnea syndrome. Sleep 1995;18:873–875.
37 Soh J, Chew MT, Wong HB: A comparative assessment of the perception of Chinese facial profile esthetics. Am J Orthod Dentofacial Orthop 2005;127:692–699.
38 Goh YH, Tan W, Mark HWI: Modified maxillomandibular advancement technique; in Friedman M (ed): Sleep Apnea and Snoring: Surgical and Non-Surgical Therapy. Philadelphia, Saunders Elsevier, 2009, pp 334–338.
39 Goh YH, Lim KA: Modified maxillomandibular advancement for the treatment of obstructive sleep apnea: a preliminary report. Laryngoscope 2003;113:1577–1582.

Prof. Dr. Sung Wan Kim
Department of Otorhinolaryngology-Head and Neck Surgery
Kyung Hee University, 23 Kyungheedae-ro, Dongdaemun-gu
Seoul 02447 (Republic of Korea)
E-Mail drkimsw@hanmail.net

Lin H-C (ed): Sleep-Related Breathing Disorders. Adv Otorhinolaryngol. Basel, Karger, 2017, vol 80, pp 109–115
DOI: 10.1159/000470879

Multilevel Obstructive Sleep Apnea Surgery

Hsin-Ching Lin[a] • Edward M. Weaver[b, c] • Ho-Sheng Lin[d] • Michael Friedman[e, f]

[a]Department of Otolaryngology, Sleep Center, Kaohsiung Chang Gung Memorial Hospital and Chang Gung University College of Medicine, Kaohsiung, Taiwan; [b]Department of Otolaryngology – Head and Neck Surgery, University of Washington, and [c]Surgery Service, Department of Veterans Affairs Medical Center, Seattle, WA, [d]Department of Otolaryngology – Head and Neck Surgery, Wayne State University School of Medicine and Karmanos Cancer Institute, Detroit Medical Center, Detroit, MI, [e]Department of Otolaryngology – Head and Neck Surgery, Division of Sleep Surgery, Rush University Medical Center, and [f]Department of Otolaryngology, Advanced Center for Specialty Care, Advocate Illinois Masonic Medical Center, Chicago, IL, USA

Abstract

Continuous positive airway pressure (CPAP) is the primary treatment of obstructive sleep apnea/hypopnea syndrome (OSA). Most sleep physicians are in agreement that a certain number of OSA patients cannot or will not use CPAP. Although other conservative therapies, such as oral appliance, sleep hygiene, and sleep positioning, may help some of these patients, there are many who fail all conservative treatments. As sleep surgeons, we have the responsibility to screen patients for both symptoms and signs of OSA. As experts of upper airway diseases, we often view an airway clearly and help the patient understand the importance of assessment and treatment for OSA. Surgery for OSA is not a substitute for CPAP but is a salvage treatment for those who failed CPAP and other conservative therapies and therefore have no other options. Most early studies and reviews focused on the efficacy of uvulopalatopharyngoplasty, a single-level procedure for the treatment of OSA. Since OSA is usually caused by multilevel obstructions, the true focus on efficacy should be on multilevel surgical intervention. The purpose here is to provide an updated overview of multilevel surgery for OSA patients.

To know the real success rate of obstructive sleep apnea (OSA) therapy, we need to be realistic about the treatment methods and the difference between efficacy and effectiveness. If the patient opts for no treatment, the success rate is essentially zero. The success rate for continuous positive airway pressure (CPAP) or oral appliance varies significantly and depends on the patient's usage. Salvage OSA surgery is a treatment for patients who fail conservative treatments. Therefore, the success rate for this group of patients can only be zero before surgery. Previous studies on the efficacy of surgery for OSA have focused on the most commonly performed procedure, uvulopalatopharyngoplasty (UPPP). A literature review of UPPP as an isolated procedure indicated that the success rate was 40.7% in nonselected patients [1]. This "40% success" figure has been misrepresented as being an indication of surgical success for OSA surgery by many sleep physicians. We all know that UPPP (and its modifications) is still currently the most common procedure for the treatment of OSA. However, if we review the first UPPP paper by Shiro Fujita [2], he mentioned that most OSA patients had a combination of more than 2 obstruction levels. Therefore, it appears that Fujita himself never intended to suggest that UPPP, single-level surgery, could cure most patients with OSA.

In 1993, Riley et al. [3] reported their surgical experience, outlining a multilevel concept. Each patient was classified as having single-level obstruction involving the oropharynx only, type 1, or the hypopharynx only, type 3. Multilevel obstruction was identified as type 2 and implied a combination of oropharyngeal and hypopharyngeal obstruction. Of the 239 patients, 93.3% (223 patients) were identified as having multilevel obstruction, type 2. Only 16 patients (6.7%) had single-level obstruction. Dr. Abdullah [4] from Hong Kong performed drug-induced sleep endoscopy (DISE) to study the obstruction level and their results showed that 87% of patients had multilevel obstruction. More recently, Ravesloot and de Vries [5] conducted a prospective study of 100 consecutive DISE procedures and investigated the associations between polysomnography and DISE findings. Their results suggest that a multilevel collapse, a complete collapse, and a tongue-base collapse are statistically significantly associated with higher apnea hypopnea index values. Vroegop et al. [6] studied a large cohort of patients with sleep-disordered breathing during DISE and found that multilevel collapse was noted in 68.2% of the 1,249 patients. The most frequently observed multilevel collapse pattern was a combination of palatal and tongue base collapse (25.5%).

We all know that the most important factor for the success of OSA surgery is patient selection. Between 2002 and 2005, Friedman et al. [7–9] published a series of papers introducing and refining the "Friedman Staging System" to help physicians select appropriate surgical candidates through examining the upper airway anatomy. Two important concepts from these papers were: (1) the anatomic basis should be the true prognosis factor for OSA surgery, not only the severity of the disease, and (2) if you add the additional tongue base surgery to the UPPP, the surgical success rate will increase [9]. This is also indirect proof that the majority of OSA patients have more than 1 level of obstruction.

The key of OSA surgery is to create a noncollapsible airspace, reduce airway resistance, and maintain adequate airflow with normal inspiratory effort when sleeping. The percentage of OSA papers on surgical intervention published on multilevel surgery has been increasing, especially in recent years. This could be due to more surgeons accepting the concept of OSA being a multilevel airway disease, which requires a multilevel treatment.

Classifications of Multilevel Surgery

Published papers on apnea-hypopnea index outcomes of multilevel surgery of the upper airway for OSA can be divided into 4 groups [10].

1. The most commonly performed multilevel approach includes a UPPP/modified UPPP as a basic technique with a second procedure designed to improve the hypopharyngeal airway. Most commonly, this includes genioglossus advancement, thyrohyoid suspension, radiofrequency tissue volume reduction/ablation of the tongue base, tongue base resection with the assistance of transoral coblation or robotic surgery, and in some cases tongue base suspension. The success rate for these procedures has been reported to be between 20 and 100%.
2. The second group of OSA patients studied who have undergone multilevel treatment include those who have had more invasive and more radical hypopharyngeal surgery, such as open tongue base resection. Because of the aggressive nature of these procedures, some of these patients would have temporary tracheotomy and require significant hospitalization. There was also significant postoperative morbidity. The success rate in this group varied between 44 and 100%.

3. The third group of multilevel surgery for OSA included those patients undergoing bimaxillary advancement, which itself is a multilevel treatment. Many of these patients had undergone a staged surgery often with UPPP and genioglossus advancement as their primary procedure. The success rate of this group was 86.0% [11].

4. The fourth group is multilevel minimally invasive techniques for OSA. Steward [12] studied 22 patients who underwent combined radiofrequency reduction of the palate and the base of tongue and reported a success rate of 59%. Fischer et al. [13] presented a similar study on multilevel minimally invasive surgery with radiofrequency on the palate, tonsils and tongue base for 15 OSA patients with a success rate of 33%. Stuck et al. [14] published their results with radiofrequency on the palate and base of the tongue for 18 OSA patients with mild/moderate disease with a success rate of 44%. In 2007, we presented minimally invasive single-stage multilevel surgery (MISS MLS) for patients with OSA [15]. Our patients had undergone 3-level treatment that included nasal surgery, palatal stiffening by a palatal implant technique, and radiofrequency volume reduction of the tongue base with a minimum follow-up of 6 months. Classical success was achieved in 54 out of 122 patients (47.5%). The snoring severity improved (reduced by at least 50%) in 77.9% of patients. Ceylan et al. [16] studied 26 OSA patients who underwent combined radiofrequency reduction of the palate and the base of tongue and reported a success rate of 53.8%. Neruntarat and Chantapant [17] also investigated 72 patients who underwent combined radiofrequency reduction of the palate and the base of tongue, and reported a success rate of 55.6%. We further studied the effects of MISS MLS on Asian patients, finding a surgical response rate of 46.8% and classical success rate of only 14.9% [18]. Our results show the improvement of the 2 relevant clinical outcomes in snoring severity and daytime sleepiness after MISS MLS for OSA patients, but the limited effects on the polysomnographic parameters in the Asian population.

Although the group of patients with minimally invasive OSA surgery had a significant failure rate, few serious complications occurred (e.g., rare abscess) and the patients who did fail were still candidates for additional traditional salvage invasive procedures. We believe that minimally invasive OSA surgery could be a viable option in the selection of OSA patients who refuse invasive airway surgery, especially in patients with a major complaint of snoring and hypersomnia.

Single- versus Multistage Surgery in Multilevel OSA Treatment

Multilevel treatment can be performed in a single stage or in multiple stages. Most studies revealed that single-stage surgery at multiple levels did not increase postoperative complications; however, these studies were too underpowered to claim equivalence with confidence [19–21]. A large cohort study [22] showed an elevated risk of serious complication when UPPP was combined with a concurrent nonnasal procedure compared to UPPP alone (adjusted risk ratio 4.8, 95% confidence interval 2.5–9.1), but no elevated risk when UPPP was combined with a nasal procedure. This study did not examine the cumulative risk of staged surgery. Single-stage multilevel surgery can lower total hospitalization expenses when compared to multi-staged surgery [20]. Kieff and Busaba [23] reported the safety of same-day or overnight discharge and concluded that same-day discharge for patients who have undergone combined nasal and palatal surgery for OSA is relatively safe in selected cases when significant comorbid diseases are not present.

Thus, on the issue of single- or multiple-stage surgery, our opinion is that it is safe and efficacious to perform multilevel surgery in 1 surgical session with adequate postoperative monitoring for selected patients with OSA. When multilevel invasive OSA procedures are necessary, the authors prefer to perform the nasal surgery without postoperative nasal packs.

Outcomes of Multilevel Obstructive Sleep Apnea Surgery

In order to apply evidence-based medicine criteria to the question of whether multilevel surgery of the upper airway may benefit patients with OSA, we performed a systematic review of all published series focusing on the subjective and objective outcomes of OSA patients treated with multilevel surgery in 2008 [10]. In our review, 49 multilevel OSA surgery articles were identified. There were 1,978 patients included in this analysis. There were significant changes in the apnea index, apnea-hypopnea index, percent of rapid eye movement sleep, minimal saturation of oxygen, snoring 0–10 visual analog scale, Epworth Sleepiness Scale, and quality of life. The original reported success rate in the included literature was 64.5%. However, the definition of success used by the authors was not consistent. Therefore, a meta-analysis was performed with the success rate defined as the postoperative apnea-hypopnea index reduced by >50% and <20 events/h, which is a commonly used criterion for success. The recalculated success rate was 66.4%. The overall complications were minor with a rate of 14.6%.

In the last decade, several surgical advances have been made in the management of OSA. There have been many new techniques or modifications of traditional surgical procedures with reported more favorable outcomes. The relatively new modifications included Z-palatopharyngoplasty, lateral pharyngoplasty, expansion sphincter pharyngoplasty, relocation pharyngoplasty, transoral endoscopic coblator tongue base resection, transoral robotic OSA surgery, and many others. Since the patient load has nearly doubled with many new publications, it is important to reassess the published success rate. We further updated the above results [8] and found that there were 43 new articles on multilevel OSA surgery up to the end of June 2016 (unpubl. data) [24–65]. A total of 87 multilevel OSA surgery articles were identified. There were 3,931 patients included in this update. According to the success rate defined above, the updated recalculated success rate is 61.6%.

This analysis, however, does not account for patient selection, confounding, and other factors that may impact on success rates. There is a lack of randomized and controlled trials in the enrolled literature. Although some of the enrolled studies reviewed here report well-controlled cohort studies, many others consist of uncontrolled case series. The limitations of meta-analyses on observational studies have been well documented, including confounding factors, selection bias, and significant heterogeneity among the published studies. The meta-analysis presented here focuses on objective outcome measurements to define treatment success. Improvement in quality of life is also important for OSA treatment and should be taken into consideration when designing and evaluating multilevel OSA therapy in the future.

Conclusion

Sleep physicians have the responsibility of helping the groups of OSA patients who cannot or will not accept CPAP as a permanent form of management. Based on the updated systematic review of multilevel OSA surgery, the true OSA surgical success rate is 61.6%, an improvement over isolated UPPP success of 40.7%. Therefore, multilevel upper airway surgery for patients with OSA is clearly associated with improved outcomes.

References

1 Sher AE, Schechtman KB, Piccirillo JF: The efficacy of surgical modifications of the upper airway in adults with obstructive sleep apnea syndrome. Sleep 1996; 19:156–177.

2 Fujita S: UPPP for sleep apnea and snoring. Ear Nose Throat J 1984;63:227–235.

3 Riley RW, Powell NB, Guilleminault C: Obstructive sleep apnea syndrome: a review of 306 consecutively treated surgical patients. Otolaryngol Head Neck Surg 1993;108:117–125.

4 Abdullah VJ, van Hasselt CA: Video sleep nasendoscopy; in Terris DJ, Goode RL (eds): Surgical Management of Sleep Apnea and Snoring. Boca Raton, Taylor and Francis, 2005, pp 143–154.

5 Ravesloot MJ, de Vries N: One hundred consecutive patients undergoing drug-induced sleep endoscopy: results and evaluation. Laryngoscope 2011;121: 2710–2716.

6 Vroegop AV, Vanderveken OM, Boudewyns AN, et al: Drug-induced sleep endoscopy in sleep-disordered breathing: report on 1,249 cases. Laryngoscope 2014;124:797–802.

7 Friedman M, Vidyasagar R, Bliznikas D, Joseph N: Does severity of obstructive of sleep apnea/hypopnea syndrome predict uvulopalatopharyngoplasty outcome? Laryngoscope 2005;115:2109–2113.

8 Friedman M, Ibrahim H, Bass L: Clinical staging for sleep-disordered breathing. Otolaryngol Head Neck Surg 2002;127: 13–21.

9 Friedman M, Ibrahim H, Lee G, Joseph NJ: Combined uvulopalatopharyngoplasty and radiofrequency tongue base reduction for treatment of obstructive sleep apnea/hypopnea syndrome. Otolaryngol Head Neck Surg 2003;129:611–621.

10 Lin HC, Friedman M, Chang HW, et al: The efficacy of multilevel surgery of the upper airway in adults with obstructive sleep apnea/hypopnea syndrome. Laryngoscope 2008;118:902–908.

11 Holty JE, Guilleminault C: Maxillomandibular advancement for the treatment of obstructive sleep apnea: a systematic review and meta-analysis. Sleep Med Rev 2010;14:287–297.

12 Steward D: Effectiveness of multilevel (tongue and palate) radiofrequency tissue ablation for patients with obstructive sleep apnea syndrome. Laryngoscope 2004;114:2073–2084.

13 Fischer Y, Khan M, Mann WJ: Multilevel temperature-controlled radiofrequency therapy of soft palate, base of tongue, and tonsils in adults with obstructive sleep apnea. Laryngoscope 2003;113: 1786–1791.

14 Stuck BA, Starzak K, Hein G, Verse T, Hormann K, Maurer JT: Combined radiofrequency surgery of the tongue base and soft palate in obstructive sleep apnoea. Acta Otolaryngol 2004;124:827–832.

15 Friedman M, Lin HC, Gurpinar B, Joseph NJ: Minimally invasive single-stage multilevel treatment for obstructive sleep apnea/hypopnea syndrome. Laryngoscope 2007;117:1859–1863.

16 Ceylan K, Emir H, Kizilkaya Z, et al: First-choice treatment in mild to moderate obstructive sleep apnea: single-stage, multilevel, temperature-controlled radiofrequency tissue volume reduction or nasal continuous positive airway pressure. Arch Otolaryngol Head Neck Surg 2009;135:915–919.

17 Neruntarat C, Chantapant S: Radiofrequency surgery for the treatment of obstructive sleep apnea: short-term and long-term results. Otolaryngol Head Neck Surg 2009;141:722–726.

18 Lin HC, Friedman M, Chang HW, Bonzelaar L, Salapatas AM, Lin MC, Huang KT: Minimally invasive single-stage multilevel surgery for OSA in Asian patients. JAMA Otolaryngol Head Neck Surg 2017;143:147–154.

19 Pang KP: One-stage nasal and multi-level pharyngeal surgery for obstructive sleep apnoea: safety and efficacy. J Laryngol Otol 2005;119:272–276.

20 Li HY, Wang PC, Hsu CY, Lee SW, Chen NH, Liu SA: Combined nasal-palatopharyngeal surgery for obstructive sleep apnea: simultaneous or staged? Acta Otolaryngol 2005;125:298–303.

21 Busaba NY: Same-stage nasal and palatopharyngeal surgery for obstructive sleep apnea: is it safe? Otolaryngol Head Neck Surg 2002;126:399–403.

22 Kezirian EJ, Weaver EM, Yueh B, Khuri SF, Daley J, Henderson WG: Risk factors for serious complication after uvulopalatopharyngoplasty. Arch Otolaryngol Head Neck Surg 2006;132:1091–1098.

23 Kieff DA, Busaba NY: Same-day discharge for selected patients undergoing combined nasal and palatal surgery for obstructive sleep apnea. Ann Otol Rhinol Laryngol 2004;113:128–131.

24 Richard W, Kox D, den Herder C, van Tinteren H, de Vries N: One stage multilevel surgery (uvulopalatopharyngoplasty, hyoid suspension, radiofrequent ablation of the tongue base with/without genioglossus advancement), in obstructive sleep apnea syndrome. Eur Arch Otorhinolaryngol 2007;264:439–444.

25 Friedman M, Lin HC, Gurpinar B, Joseph NJ: Minimally invasive single-stage multilevel treatment for obstructive sleep apnea/hypopnea syndrome. Laryngoscope 2007;117:1859–1863.

26 Yin SK, Yi HL, Lu WY, Guan J, Wu HM, Cao ZY: Genioglossus advancement and hyoid suspension plus uvulopalatopharyngoplasty for severe OSAHS. Otolaryngol Head Neck Surg 2007;136:626–631.

27 Friedman M, Soans R, Gurpinar B, Lin HC, Joseph N: Evaluation of submucosal minimally invasive lingual excision technique for treatment of obstructive sleep apnea/hypopnea syndrome. Otolaryngol Head Neck Surg 2008;139:378–384.

28 van den Broek E, Richard W, van Tinteren H, de Vries N: UPPP combined with radiofrequency thermotherapy of the tongue base for the treatment of obstructive sleep apnea syndrome. Eur Arch Otorhinolaryngol 2008;265:1361–1365.

29 Eun YG, Kim SW, Kwon KH, Byun JY, Lee KH: Single-session radiofrequency tongue base reduction combined with uvulopalatopharyngoplasty for obstructive sleep apnea syndrome. Eur Arch Otorhinolaryngol 2008;265:1495–1500.

30 Lye KW, Waite PD, Meara D, Wang D: Quality of life evaluation of maxillomandibular advancement surgery for treatment of obstructive sleep apnea. J Oral Maxillofac Surg 2008;66:968–972.

31 Eun YG, Kwon KH, Shin SY, Lee KH, Byun JY, Kim SW: Multilevel surgery in patients with rapid eye movement-related obstructive sleep apnea. Otolaryngol Head Neck Surg 2009;140:536–541.

32 Babademez MA, Ciftci B, Acar B, et al: Low-temperature bipolar radiofrequency ablation (coblation) of the tongue base for supine-position-associated obstructive sleep apnea. ORL J Otorhinolaryngol Relat Spec 2010;72:515–515.

33 Kezirian EJ, Malhotra A, Goldberg AN, White DP: Changes in obstructive sleep apnea severity, biomarkers, and quality of life after multilevel surgery. Laryngoscope 2010;120:1481–1488.

34 Jones R, Badlani J, Jones C: Maxillary, mandibular and chin advancement surgery for the treatment of obstructive sleep apnoea. Aust Dent J 2010;55:314–321.

35 Vicini C, Dallan I, Canzi P, Frassineti S, La Pietra MG, Montevecchi F: Transoral robotic tongue base resection in obstructive sleep apnoea-hypopnoea syndrome: a preliminary report. ORL J Otorhinolaryngol Relat Spec 2010;72:22–27.

36 Yaremchuk K, Tacia B, Peterson E, Roth T: Change in Epworth Sleepiness Scale after surgical treatment of obstructive sleep apnea. Laryngoscope 2011;121:1590–1593.

37 Emara TA, Omara TA, Shouman WM: Modified genioglossus advancement and uvulopalatopharyngoplasty in patients with obstructive sleep apnea. Otolaryngol Head Neck Surg 2011;145:865–871.

38 Koutsourelakis I, Safiruddin F, Ravesloot M, Zakynthinos S, de Vries N: Surgery for obstructive sleep apnea: sleep endoscopy determinants of outcome. Laryngoscope 2012;122:2587–2591.

39 Suh GD: Evaluation of open midline glossectomy in the multilevel surgical management of obstructive sleep apnea syndrome. Otolaryngol Head Neck Surg 2013;148:166–171.

40 Kim SJ, Kim YS, Park JH, Kim SW: Cephalometric predictors of therapeutic response to multilevel surgery in patients with obstructive sleep apnea. J Oral Maxillofac Surg 2012;70:1404–1412.

41 Lee JM, Weinstein GS, O'Malley BW Jr, Thaler ER: Transoral robot-assisted lingual tonsillectomy and uvulopalatopharyngoplasty for obstructive sleep apnea. Ann Otol Rhinol Laryngol 2012;121:635–639.

42 Stow NW, Sale PJ, Lee D, Joffe D, Gallagher RM: Simultaneous tonsillectomy and nasal surgery in adult obstructive sleep apnea: a pilot study. Otolaryngol Head Neck Surg 2012;147:387–391.

43 Vicini C, Dallan I, Canzi P, et al: Transoral robotic surgery of the tongue base in obstructive sleep apnea-hypopnea syndrome: anatomic considerations and clinical experience. Head Neck 2012;34:15–22.

44 Friedman M, Hamilton C, Samuelson CG, Kelley K, Taylor D, Pearson-Chauhan K, Maley A, Taylor R, Venkatesan TK: Transoral robotic glossectomy for the treatment of obstructive sleep apnea-hypopnea syndrome. Otolaryngol Head Neck Surg 2012;146:854–862.

45 Plzak J, Zabrodsky M, Kastner J, Betka J, Klozar J: Combined bipolar radiofrequency surgery of the tongue base and uvulopalatopharyngoplasty for obstructive sleep apnea. Arch Med Sci 2013;9:1097–1101.

46 MacKay SG, Carney AS, Woods C, et al: Modified uvulopalatopharyngoplasty and coblation channeling of the tongue for obstructive sleep apnea: a multi-centre Australian trial. J Clin Sleep Med 2013;9:117–124.

47 Gunawardena I, Robinson S, MacKay S, Woods CM, Choo J, Esterman A, Carney AS: Submucosal lingualplasty for adult obstructive sleep apnea. Otolaryngol Head Neck Surg 2013;148:157–165.

48 MacKay SG, Jefferson N, Grundy L, Lewis R: Coblation-assisted Lewis and MacKay operation (CobLAMO): new technique for tongue reduction in sleep apnoea surgery. J Laryngol Otol 2013;127:1222–1225.

49 Lin HS, Rowley JA, Badr MS, et al: Transoral robotic surgery for treatment of obstructive sleep apnea-hypopnea syndrome. Laryngoscope 2013;123:1811–1816.

50 Toh ST, Han HJ, Tay HN, Kiong KL: Transoral robotic surgery for obstructive sleep apnea in Asian patients: a Singapore sleep centre experience. JAMA Otolaryngol Head Neck Surg 2014;140:624–629.

51 van Maanen JP, Witte BI, de Vries N: Theoretical approach towards increasing effectiveness of palatal surgery in obstructive sleep apnea: role for concomitant positional therapy? Sleep Breath 2014;18:341–349.

52 Lin HC, Friedman M, Chang HW, Yalamanchali S: Z-palatopharyngoplasty combined with endoscopic coblator open tongue base resection for severe obstructive sleep apnea/hypopnea syndrome. Otolaryngol Head Neck Surg 2014;150:1078–1085.

53 Vicini C, Montevecchi F, Campanini A, et al: Clinical outcomes and complications associated with TORS for OSAHS: a benchmark for evaluating an emerging surgical technology in a targeted application for benign disease. ORL J Otorhinolaryngol Relat Spec 2014;76:63–69.

54 Vicini C, Montevecchi F, Pang K, Bahgat A, Dallan I, Frassineti S, Campanini A: Combined transoral robotic tongue base surgery and palate surgery in obstructive sleep apnea-hypopnea syndrome: expansion sphincter pharyngoplasty versus uvulopalatopharyngoplasty. Head Neck 2014;36:77–83.

55 Hoff PT, Glazer TA, Spector ME: Body mass index predicts success in patients undergoing transoral robotic surgery for obstructive sleep apnea. ORL J Otorhinolaryngol Relat Spec 2014;76:266–272.

56 Chen S, Shi S, Xia Y, Zhu M, Zhang C, Xia S, Zheng H: A prospective study of the surgical outcome of simple uvulopalatopharyngoplasty (UPPP), UPPP combined with genioglossus advancement or tongue base advancement for obstructive sleep apnea hypopnea syndrome patients with multilevel obstruction. Clin Exp Otorhinolaryngol 2015;8:136–141.

57 Verse T, Wenzel S, Brus J: Multi-level surgery for obstructive sleep apnea: lingual tonsillectomy vs. hyoid suspension in combination with radiofrequency of the tongue base. Sleep Breath 2015;19:1361–1366.

58 Turhan M, Bostanci A, Bozkurt S: Predicting the outcome of modified tongue base suspension combined with uvulopalatopharyngoplasty. Eur Arch Otorhinolaryngol 2015;272:3411–3416.

59 Zhang XM, Tham CJ, Yin YL, Sun YQ, Zhou X: A novel palatal implant surgery combined with uvulopalatopharyngoplasty and inferior turbinate radiofrequency for the treatment of moderate to severe obstructive sleep apnea: a pilot study. Eur Arch Otorhinolaryngol 2015;272:1195–1202.

60 Bostanci A, Turhan M: A systematic review of tongue base suspension techniques as an isolated procedure or combined with uvulopalatopharyngoplasty in obstructive sleep apnea. Eur Arch Otorhinolaryngol 2016;273:2895–2901.
61 Lin HS, Rowley JA, Folbe AJ, Yoo GH, Badr MS, Chen W: Transoral robotic surgery for treatment of obstructive sleep apnea: factors predicting surgical response. Laryngoscope 2015;125:1013–1020.
62 Chiffer RC, Schwab RJ, Keenan BT, Borek RC, Thaler ER: Volumetric MRI analysis pre- and post-transoral robotic surgery for obstructive sleep apnea. Laryngoscope 2015;125:1988–1995.
63 Yüksel A, Ugur KS, Kizilbulut G, Ark N, Kurtaran H, Kaya M, Gunduz M: Long-term results of one staged multilevel surgery with tongue suspension surgery or one level palatal surgery for treatment of moderate and severe obstructive sleep apnea. Eur Arch Otorhinolaryngol 2016;273:1227–1234.
64 Thaler ER, Rassekh CH, Lee JM, Weinstein GS, O'Malley BW Jr: Outcomes for multilevel surgery for sleep apnea: obstructive sleep apnea, transoral robotic surgery, and uvulopalatopharyngoplasty. Laryngoscope 2016;126:266–269.
65 Li HY, Lee LA, Kezirian EJ: Coblation endoscopic lingual lightening (CELL) for obstructive sleep apnea. Eur Arch Otorhinolaryngol 2016;273:231–236.

Hsin-Ching Lin, MD
Department of Otolaryngology, Sleep Center
Kaohsiung Chang Gung Memorial Hospital, 123, Ta-Pei Road
Niao-Sung District, Kaohsiung City 833, Taiwan (ROC)
E-Mail enthclin@aol.com

Lin H-C (ed): Sleep-Related Breathing Disorders. Adv Otorhinolaryngol. Basel, Karger, 2017, vol 80, pp 116–124
DOI: 10.1159/000470880

Innovative Surgery for Obstructive Sleep Apnea: Nerve Stimulator

J. Ulrich Sommer · Karl Hörmann

Department of Otorhinolaryngology, Head and Neck Surgery, University Hospital Mannheim, Mannheim, Germany

Abstract

Obstructive sleep apnea (OSA) as a multifactorial disease is treated with continuous positive airway pressure (CPAP) as the gold standard. Yet, if patients suffer from CPAP incompliance, traditional OSA surgery only targets morphological changes of the upper airway while neglecting functional issues. With the advent of upper airway stimulation, and in particular hypoglossal nerve stimulation as a treatment option, a highly effective, clinically proven and functional therapy with good evidence is available. This article gives a comprehensive overview of current and upcoming hypoglossal nerve stimulation systems (Inspire, ImThera, and Nyxoah), the specific advantages of this approach, the selection criteria and screening process, relevant clinical data, and a description of the different implantation procedures. Upper airway stimulation and hypoglossal nerve stimulation appears to be a long-term, low-morbidity treatment for moderate-to-severe OSA patients suffering from CPAP incompliance.

The mechanisms causing an upper airway collapse have not been completely understood up until now. Yet, a decline in pharyngeal neuromuscular activity during sleep in obstructive sleep apnea (OSA) patients has been identified. This knowledge has supported the notion that stimulation of the upper airway muscles may prove effective in treating this condition. Previous reports have indicated that various upper airway dilator muscles, particularly the genioglossus, contribute to stabilizing the upper airway during sleep. As a result, various actions have been taken to selectively stimulate only the upper airway dilator muscles. This has been achieved using transcutaneous, percutaneous, and transmucosal stimulation of the hypoglossal nerve. The next step in the evolution of neuromodulation for upper airway stimulation was the development of implantable neurostimulation units consisting of an implantable pulse generator

(IPG) and a cuffed electrode wrapped around the hypoglossal nerve [1].

Since the advent of these IPGs as treatment for OSA, quite some time has passed and the their use has become more widespread around the world. Following the publication of the STAR trial data in the New England Journal of Medicine, the patients treated in that group remained under control and in 2016 the 36-month follow-up data were published. These data show that treatment efficiency and patient compliance remained constantly high over the 3 years. From the 3 systems that were initially commercially available, only the Inspire and the ImThera are still offered. Other systems, such as the Nyxoah, have so far only been used in safety and performance studies [2, 3].

With the development of upper airway stimulation, the operative arsenal in surgery for sleep-disordered breathing has vastly increased. The therapeutic principle behind all the systems is an electrical stimulation of the hypoglossal nerve, which leads to an increase in the genioglossal muscle tone, and a slight protrusion of the tongue. This movement and the tonification opens the upper airway and prevents a collapse through stabilization. Currently, all available systems are recommended for patients with medium to severe OSA and only mild to medium obesity.

Anatomy

Stimulation of the upper airway dilator muscles is achieved through transcutaneous, percutaneous, and transmucosal stimulation of the hypoglossal nerve. In humans, the lateral hypoglossal nerve mainly innervates the hyoglossus and styloglossus muscles, whereas the medial hypoglossal nerve innervates the genioglossus muscle. The genioglossus muscle represents the main tongue protrusor. The hyoglossus and styloglossus muscle are mainly retrusors and the intrinsic muscles are involved in modifying the shape of the tongue. In most patients, the lateral branch of the hypoglossal nerve consists of multiple branches. The medial branch divides into its terminal fibers before entering the genioglossus muscle laterally. Therefore, a stimulation of the main trunk or 1 of the major divisions will provoke a coactivation of extrinsic and intrinsic tongue muscles [1].

Preoperative Evaluation

The preoperative evaluation of patients should include a polysomnography. As the clinical data suggest, patients should not be obese (BMI ≤35, or probably better ≤32) and have an apnea-hypopnea index (AHI) between 15 and 65 (probably better ≤50). A positional dependency as well as a high number of central or mixed apneas seems unfavorable.

Contraindications

The most important contraindications for implantation of an upper airway stimulation system include hypoglossal nerve palsy, neuromuscular diseases, radiotherapy or ablation therapy of the head and/or neck, surgical resection for cancer or congenital malformations in the larynx, tongue, or throat (with the exception of tonsillectomy and/or adenoidectomy), obvious fixed upper airway obstructions (tumors, polyps, nasal obstruction), or anticipated exposure to magnetic resonance imaging, radiation therapy, or diathermy treatment.

The Inspire System

Stimulation Device

The Inspire IPG is implanted subcutaneously in a pocket. This pocket is formed caudal of the right clavicle. It features 2 connecting jacks: 1 for the stimulation lead and 1 for the sensing lead. The IPG contains a nonrechargeable battery (which,

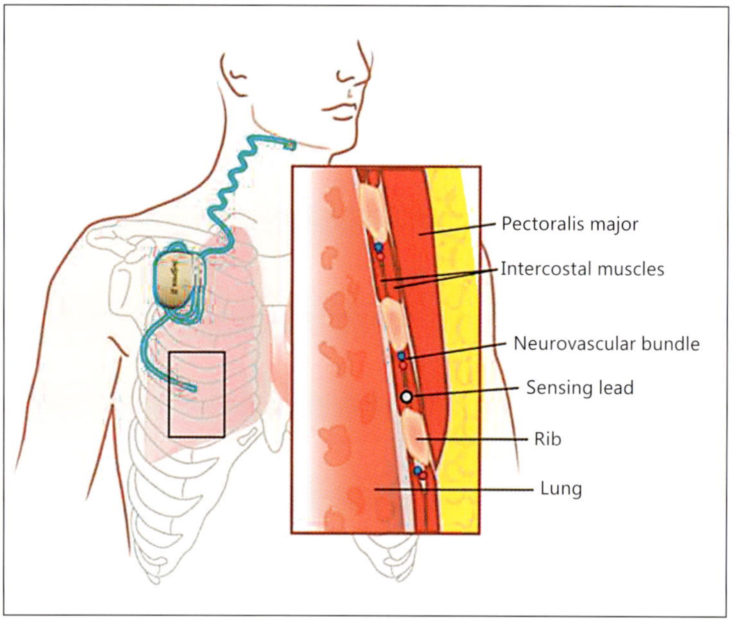

Fig. 1. Components of the Inspire system showing the stimulation electrode, the implantable pulse generator, and the sensing lead. The box displays the sensing lead placed between the intercostal muscles.

according to latest calculations, lasts for a mean duration of 10.6 years), embodied in a titanium case. The signal-processing unit is able to synchronize the hypoglossal nerve stimulation with the respiration signal. This synchronization may be adapted, inverted, and modified in many ways to suit for the individual patient.

The stimulation lead features a standard, ISO-certified, IS-1 low-profile implantable pacemaker connector to link the IPG to the cuff electrode. The cuffed electrode is made of soft silicone, and includes 3 electrodes. The outer 2 electrodes are electrically connected. Usually, the center electrode is connected to the cathode of the stimulation and the outer electrodes are connected to the anode of the stimulation system. Yet, other configurations, such as unipolar stimulation using the titanium case as a ground electrode, are possible by programming the IPG afterwards. Also, the sensing lead is connected to the IPG with an IS-1 connector. In this way, a differential pressure sensor transmits the intrathoracic pressure changes to the IPG.

The device is shipped with a small remote control in order to activate the system at night, and terminate or pause stimulation. When one decides to launch the stimulation a start delay is activated. The remote control communicates with the IPG utilizing a short-range radiofrequency signal. The patient places the programmer remote over the implanted IPG site, and uses push buttons on top of it to turn the therapy on or off, temporarily suspend the therapy, or make adjustments to the stimulation amplitude (volts) within the limits set by the physician. The remote also allows patients to check the status of the IPG battery. Figure 1 provides a system components overview.

Stimulation Pattern

Analog to continuous positive airway pressure (CPAP) and other neurostimulation therapies, the Inspire system allows adjustments in order to optimize therapy response and patient comfort. For example, the sensing parameters are adjusted by the physician to deliver the stimulation during the late expiratory phase and throughout the inspiratory phase of respiration. Furthermore, stimulation parameters like amplitude, rate, pulse width, and electrode configuration are also set by the physician during the titration nights, to provide a stable airway without waking up the patient or disturbing sleep. Adjustments of pacing and sensing parameters are done with a physician programmer unit, also using radiofrequency telemetry for communication with the IPG.

Patient Selection

A proper patient selection is crucial for success with the Inspire system. Patients must be examined while awake using endoscopy to identify anatomic factors that might obstruct the upper airway and to check for functional deficits that might hinder stimulation therapy. Using the common VOTE (velum/oropharynx/tongue base/epiglottis) classification, a drug-induced sleep endoscopy (DISE) has to be performed to rule out a complete concentric collapse at the palatal level, which is a contraindication for use. Patients with an unaffected tongue base and a sole collapse of the palate may be eligible as long as the palatal collapse is not complete and concentric [4, 5].

Surgical Implantation Procedure

After transnasal intubation and transoral placement of nerve integrity electrodes in the genioglossus and styloglossus muscles, a transparent surgical foil is draped over the mouth of the patient, which is kept open by a gauze roll placed between the teeth. The implantation procedure starts with the creation of a 4- to 6-cm-long horizontal upper neck incision on the right side, in a cervical relaxed skin tension line at least 2–3 cm below the inferior border of the mandible. The medial end of the incision should be about 0.5 cm lateral to the midline. The submandibular gland is approached, lifted, and followed down to the intermediate tendon of the digastric muscle. The main trunk of the hypoglossal nerve can then be identified by retracting the posterior border of the mylohyoid muscle. The main trunk of the hypoglossal nerve has to be dissected until the branching into the lateral (l–XII; which takes off superiorly) and medial divisions (m-XII; which runs anteriorly) at the anterior border of the hyoglossus muscle is exposed. The m-XII branch innervates the intrinsic muscles of the tongue and the genioglossus muscle, which is the principal protrusor of the tongue and is therefore the target for stimulation. Identification of the protruding branches is supported by the use of a nerve integrity monitoring system. After further dissection of the medial branch at a length of at least 1 cm, the cuff electrode is wrapped around and the effect of stimulation is verified by using the physician programmer to assess functional response. Finally, the cuff-cable-anchor is attached to the digastric tendon with 2 permanent sutures (Fig. 2) [3].

An anterior axillary incision is made to create a subcutaneous pocket which will later accommodate the IPG in the right mid-infraclavicular region superficial to the fascia of the pectoralis major muscle. The stimulation lead is subcutaneously tunneled into the IPG pocket with the supplied tunneling device. To place the sensing lead, a third 5-cm-long horizontal incision is made at the 4th to 6th intercostal space with the lateral edge anterior to the midaxillary line. When the external oblique muscle is reached, a smaller tunnel of 3–5 cm between the external and internal intercostal muscle is created using a blunt instrument. The pressure sensor is placed in this tunnel, with the sensing (curved) side directed towards the lung. Two anchors fix the sensor with the cable forming an omega shape on the underlying fascia. Finally, the sensing lead is also tunneled subcutaneously back to the IPG pocket to be connected.

Before closure of the wounds, the physician programmer is used to check the respiration signal. The function of the entire system is verified and the voltage level of the first functional tongue movements is recorded. Special attention should be given to the movement pattern of the tongue. Best results are achieved by having a clear protru-

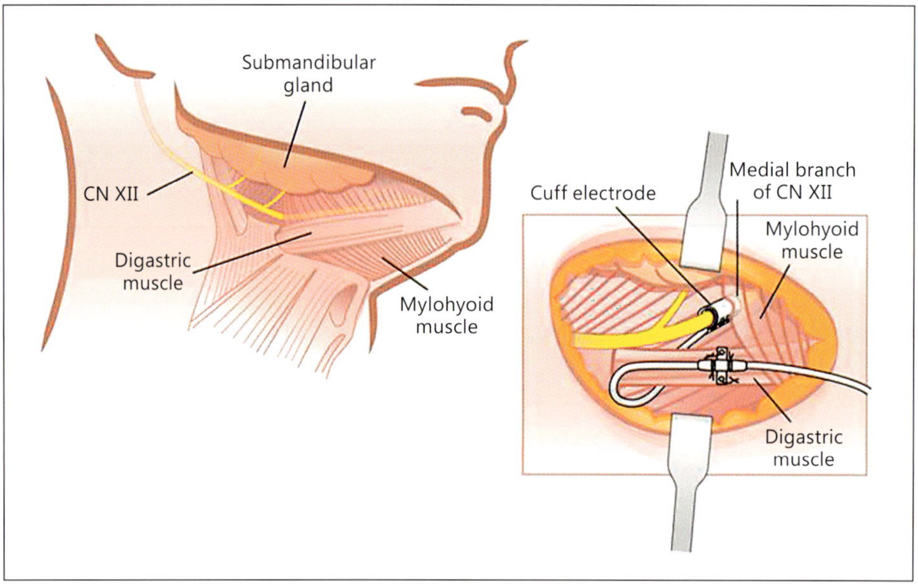

Fig. 2. The Inspire stimulation lead placed around a distal branch of the hypoglossal nerve.

sion without retracting components. If protrusion appears not to be adequate, the cuff should be repositioned. Drainage is not recommended in order to reduce the potential for infectious contamination of the implanted components.

Clinical Data

As Inspire is the commercial system with the longest availability on the market, it also boasts the most clinical data. Eight patients were implanted in the "Inspire 1" trial, which represents the "first in human" and "proof of concept" study. The "Inspire 2" trial focused on patient selection; 22 patients were implanted in 4 different centers. Both the Inspire 2 and 3 (a total of 12 patients with implants in 7 centers) trials were also safety and efficacy studies followed by the well-known "Inspire 4" or STAR trial. The STAR trial was conducted in 22 centers; 126 patients were implanted in the study (83% were men). The median AHI score at 12 months decreased by 68%, from 29.3 to 9.0 ($p < 0.001$); the oxygen desaturation index score decreased by 70%, from 25.4 to 7.4 ($p < 0.001$). The secondary outcome measures showed a reduction in the effects of sleep apnea and improved quality of life. The STAR trial proved that upper airway stimulation can lead to significant improvements in objective and subjective measurements of the severity of OSA [2].

The implanted patient cohort remained under clinical control and the 36-month follow-up data were published recently. In this study, 81% of the patients treated with the Inspire system still used the device on a regular basis for at least 7 h per night.

A large percentage of the participants achieved normalization of clinical outcomes at 3 years. Normalization of sleepiness (Epworth Sleepiness Scale) increased from 33 to 77%, while the percentage of participants reporting a normal sleep-related quality of life (FOSQ) increased from 15 to 63% from baseline to 36 months. Finally, AHI was not only stable but even improved compared to the 12-month data.

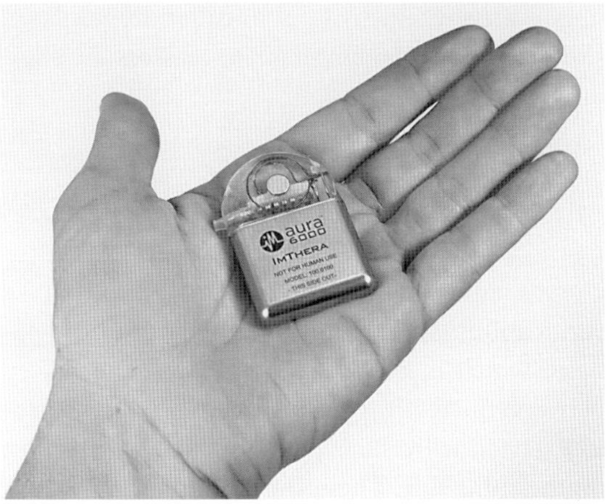

Fig. 3. The implantable pulse generator of the ImThera aura6000 system.

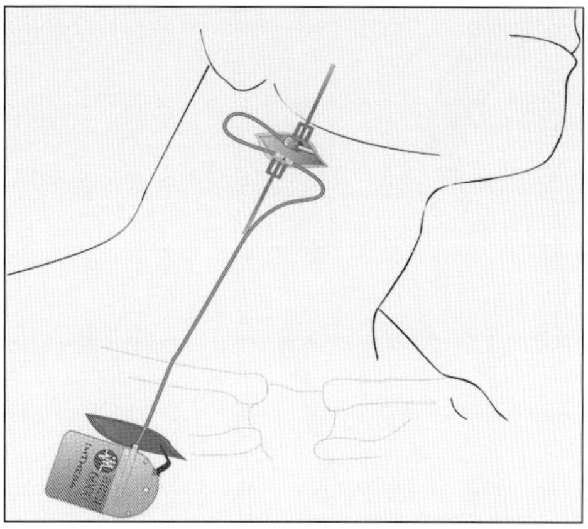

Fig. 4. Components of the ImThera system showing the stimulation electrode and the implantable pulse generator.

The only serious device-related adverse effects after 36 months included 3 device explants: 2 were at the participants' request due to other sleep disorders, and 1 was in a therapy responder that needed to be explanted due to an unrelated systemic septic arthritis. Another interesting fact is that there were no changes in stimulation threshold parameters. This indicates that fibrosis or nerve-related changes are not to be expected when using cuffed electrodes for upper airway stimulation therapy. The STAR trial therefore provides the strongest data to date that supports neuromodulation as durable and effective therapy for OSA [3].

The ImThera "THN aura6000" System

Stimulation Device

The aura6000 system consists of an IPG (Fig. 3), an implant containing the battery and stimulation system, and a cuff electrode system composed of an 8-mm soft silicone cuff housing 6 independent electrodes that are connected to the IPG via a subcutaneously tunneled lead wire. For better handling and passing under the hypoglossal nerve, the cuff electrode is equipped with 2 colored loops. The IPG battery is rechargeable. Figure 4 offers a system overview. The recharging is performed transcutaneously with an external remote control charger and a charging coil that is placed over the IPG with the help of 2 magnets. Daily charging of the device is needed and takes approximately 1 h. The same remote control charger is used to start, pause, and stop each nightly session of stimulation. The IPG contains an integrated log memory. When the patient returns to his or her physician for follow-up, the data may provide information about charging times and usage [6, 7].

Stimulation Pattern

As the aura6000 system consists of a 6-contact cuff electrode positioned around the proximal trunk of the hypoglossal nerve and does not incorporate a sensing system of the respiration, the stimulation is quite different from the Inspire device as it stimulates successively during both inspiration and expiration via selected contacts in order to induce appropriate pharyngeal and oral tongue displacements [7].

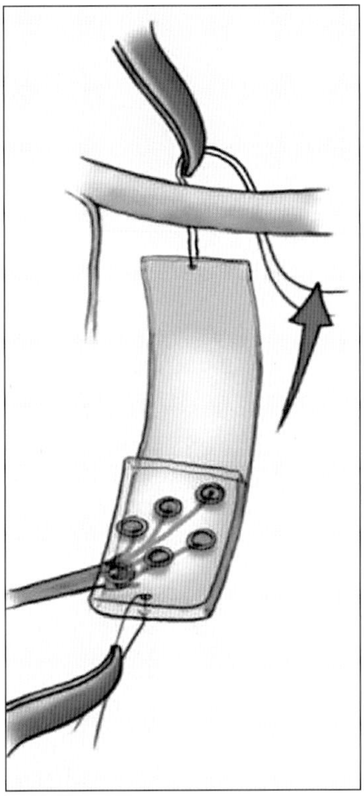

Fig. 5. The ImThera aura6000 stimulation electrode passes under the main trunk of the hypoglossal nerve. The 6 electrical contacts are oriented in 2 rows and have individual leads connecting them to the implantable pulse generator.

Patient Selection

Patient selection for the ImThera aura6000 system seems to be easier than for the Inspire system as no DISE is required during the screening process. As long as the patient has a moderate to severe OSA, has failed or does not tolerate CPAP therapy, has failed or refuses alternative OSA treatments (e.g., surgery, oral appliances, or behavioral treatments), and is not implanted with another active implantable device or has a BMI ≥35, they are theoretically suitable to be implanted with the ImThera system. Practically, a relevant number central apneas, facial malformations, or a relevant anatomical obstruction of the upper airway, such as a tonsillar hyperplasia, should prompt the physician to reassess the eligibility of the patient.

Surgical Implantation Procedure

After transnasal intubation and transoral placement of nerve integrity electrodes in the genioglossus and styloglossus muscle, a transparent surgical foil is draped over the mouth of the patient, which is kept open by a gauze roll placed between the teeth. The surgical procedure is usually started with the hypoglossal nerve dissection on the right side. Then, the mylohyoid muscle is approached at the medial part of the submandibular gland. After lifting it up, the proximal part of the hypoglossal nerve is dissected and the electrode cuff is placed around the hypoglossal nerve near the middle tendon of the digastric muscle so that the 6 stimulating electrodes are radially in contact with the cylindrical body of the proximal hypoglossal nerve (Fig. 5). Afterwards, the electrode is looped and anchored nearby. After creating an incision above the pectoral muscle, the subcutaneous pocket is created and the IPG is implanted. Using a tunneling device, a subcutaneous tunnel is created between the 2 incisions. The lead is passed downwards and connected to the IPG. After checking the system's electrical integrity, the incisions are closed. Similar to the Inspire system, drainage is not recommended in order to reduce the potential for infectious contamination of the implanted components [6, 7].

Clinical Data

Compared to the huge number of patients treated with the Inspire system, data regarding the ImThera system still falls short. The study conducted so far was published in 2013 and it showed 13 out of 14 patients were implanted successfully. At 12 months, the AHI had decreased from 45 ± 18 to 21 ± 17. This corresponds to a 53% AHI reduction ($p = 0.001$). The 4% oxygen desaturation index fell from 29 ± 20 to 15 ± 16 and the arousal index from 37 ± 13 to 25 ± 14 (both $p = 0.001$).

The Epworth Sleepiness Scale decreased from 11 ± 7 to 8 ± 4 ($p = 0.09$). Stimulation with the ImThera system was neither painful nor awakened the patients [6].

After completion of this study it was found that when the patients stopped using the ImThera system, the symptoms of OSA (including loud snoring) did not reappear for several nights. Therefore, the possible residual effect was tested in a clinical study. In contrast to CPAP treatment, both, the on- and the off-stimulation results were significantly better than the baseline results [7].

The Nyxoah System

Implantable Pulse Generator

The design of the Nyxoah sleep apnea therapy system differs dramatically from the Inspire and ImThera systems in its lack of complexity. The Nyxoah system does not have connecting wires and requires a much simpler surgery as it does not need the hypoglossal nerve to be isolated. The implantable portion of the system is completely passive and is powered externally. A tiny device, encased in soft silicon, equipped with paddle electrodes mounted on flexible and adjustable legs is implanted on top of the genioglossus muscle. The device is externally powered through a coil that converts the electromagnetic field generated by the external unit so it may deliver bilateral stimuli directly to the distal end of the hypoglossal nerve.

The external activation chip (AC), the "brain" of the system (composed of a battery and electronics), is kept in place over the implant under the chin using a disposable patch. If the patient wants to start the therapy when going to bed, he or she simply wears a patch, with the AC attached to it, under the chin. The stimulation begins after a certain delay, which is set by the physician. A change of the stimulation parameters, amplitude, or delay time is only possible by using the physician programmer. The AC also features a log file, which allows the physician to access important information about use time, battery status, and technical problems.

Patient Selection

As in the Inspire system, the Nyxoah system selectively stimulates only certain parts of the hypoglossal nerve. A proper patient selection is therefore crucial for effective treatment. Besides an examination to rule out anatomic factors obstructing the upper airway, a DISE procedure is performed using the VOTE classification. A complete concentric collapse at the palatal level is a contraindication. Similar to the Inspire patient selection criteria, patients with an unaffected tongue base and a sole collapse of the palate may be eligible as long as the palatal collapse is not complete and concentric.

Surgical Implantation Procedure

After transnasal intubation and transoral placement of nerve integrity electrodes in the genioglossus and styloglossus muscle, a transparent surgical foil is draped over the mouth of the patient, which is kept open by a gauze roll placed between the teeth. The terminal fibers of the hypoglossal nerve are reached using a medial, 4- to 6-cm-long incision parallel to the hyoid bone. After dissection of the platysma, the mylohyoid muscle is incised while strictly staying in the midline. When reaching the geniohyoid muscle, it is separated into its 2 bellies and the genioglossus muscle is exposed. By bluntly dissecting the tissue at the genioglossal muscle bellies, the terminal fibers of the hypoglossal nerve are exposed and the implant is placed on top of the genioglossus muscle (like "a saddle on a horse") with the electrode parts connected to the nerve terminal fibers. The implant is then sutured in place and the wound is closed without drainage to reduce the potential for infectious contamination of the implant.

Clinical Data

Following extensive testing in porcine models, Nyxoah devices were used for "first in human" safety studies in 2013. The "proof of concept" trial followed in 2015 showing very promising results. As the study is ongoing, citable clinical data are not yet available.

Conclusion

As sleep surgery only features high-evidence clinical studies for certain selected procedures, which are suitable only for a very limited population, treatment of CPAP-incompliant OSA patients still seems challenging. This is where, as the experience with neurostimulation therapy increases, hypoglossal nerve stimulation increasingly proves to be a valuable component of the therapeutic arsenal of OSA therapy. In contrast to upper airway surgery, for which long-term results still seem at least unclear, therapeutic neuromodulation results appear to be stable over time.

However, not every patient is suitable for implantation of a hypoglossal stimulation system. Special attention has to be given to the screening process as certain findings in the clinical examination, the BMI, or the DISE procedure will rule out certain therapies for the patient [8–10].

References

1 Bassiri Gharb B, et al: Microsurgical anatomy of the terminal hypoglossal nerve relevant for neurostimulation in obstructive sleep apnea. Neuromodulation 2015;18:721–728.

2 Strollo PJ Jr, et al: Upper-airway stimulation for obstructive sleep apnea. N Engl J Med 2014;370:139–149.

3 Woodson BT, et al: Three-year outcomes of cranial nerve stimulation for obstructive sleep apnea: the STAR Trial. Otolaryngol Head Neck Surg 2016;154:181–188.

4 Kezirian EJ, Hohenhorst W, de Vries N: Drug-induced sleep endoscopy: the VOTE classification. Eur Arch Otorhinolaryngol 2011;268:1233–1236.

5 Vanderveken OM, et al: Evaluation of drug-induced sleep endoscopy as a patient selection tool for implanted upper airway stimulation for obstructive sleep apnea. J Clin Sleep Med 2013;9:433–438.

6 Mwenge GB, et al: Targeted hypoglossal neurostimulation for obstructive sleep apnoea: a 1-year pilot study. Eur Respir J 2013;41:360–367.

7 Rodenstein D, et al: Residual effect of THN hypoglossal stimulation in obstructive sleep apnea: a disease-modifying therapy. Am J Respir Crit Care Med 2013;187:1276–1278.

8 Safiruddin F, et al: Effect of upper-airway stimulation for obstructive sleep apnoea on airway dimensions. Eur Respir J 2015;45:129–138.

9 Sommer UJ, et al: Tonsillectomy with uvulopalatopharyngoplasty in obstructive sleep apnea. Dtsch Arztebl Int 2016;113:1–8.

10 Quan SF: Treatment of obstructive sleep apnea and hypopnea – we are not there yet! J Clin Sleep Med 2009;5:189–190.

Prof. Dr. med. J. Ulrich Sommer
Department of Otorhinolaryngology
Head and Neck Surgery
University Hospital Mannheim
Theodor-Kutzer-Ufer 1-3
DE–68167 Mannheim (Germany)
E-Mail ulrich.sommer@medma.uni-heidelberg.de

Lin H-C (ed): Sleep-Related Breathing Disorders. Adv Otorhinolaryngol. Basel, Karger, 2017, vol 80, pp 125–135
DOI: 10.1159/000470882

Robotic Obstructive Sleep Apnea Surgery

Song Tar Toh[a–d] • Pon Poh Hsu[b–g]

[a]Singapore General Hospital, [b]SingHealth Duke-NUS Sleep Centre, [c]Duke-NUS Medical School, [d]School of Medicine, National University of Singapore, [e]Department of Otolaryngology, Head and Neck Surgery and [f]ENT Specialist Clinic, Changi General Hospital, and [g]Singapore University of Technology and Design, Singapore, Singapore

Abstract

Since the first report of the use of the da Vinci robotic system (Intuitive Surgical, Sunnyvale, CA, USA) in transoral robotic tongue base reduction for obstructive sleep apnea (OSA) was published in 2010, this surgical tool and technique has been used worldwide for the resection of tongue base tissue in the multilevel surgical treatment of OSA. The combined knowledge of the published literature on its use has enlightened sleep surgeons worldwide on this new yet evolving surgical tool. Here we will discuss the use of the da Vinci robotic system in the treatment of OSA, the pertinent surgical anatomy for a safe surgical procedure, the primary and secondary outcomes to expect in the multilevel and primary use of this technology in treating the tongue base, the predictors for success or failure, and the complications associated with this technique. We will also compare this technology with other existing techniques for treating OSA and look to the future for other similar technologies in this application.

Obstructive sleep apnea (OSA) is a multifactorial and complex disease associated with repeated upper airway collapse during sleep, resulting in oxygen desaturations, frequent arousals, and sleep fragmentation. Untreated OSA has been shown to have significant adverse effects on cardiovascular, cerebrovascular, metabolic, and neurocognitive functions, and contribute to poor quality of life and an increased risk of death [1–4].

Positive airway pressure (PAP) is the gold standard for treating OSA. However, PAP therapy is frequently limited by poor long-term compliance and adherence, and patients often do not want to use it [5, 6]. Surgical treatment can then be offered as an option.

Palatal surgery alone is ineffective in normalizing the apnea-hypopnea index (AHI) [7, 8]. Considering that OSA is characterized by multilevel airway collapse, multilevel surgery yields better outcomes [9].

Studies using drug-induced sleep endoscopy in the preoperative assessment of the upper airway showed that the base of the tongue and supraglottic regions are key sites of obstruction in a significant proportion of patients [10, 11]. The aim of hypopharyngeal OSA surgery is to widen the retroglossal and hypopharyngeal airway space.

Over the years, many tongue base procedures have been reported, including volumetric tongue base reduction procedures using extensive transcervical approaches to the tongue base area, radiofrequency reduction, midline glossectomy, lingual plasty, and submucosal minimally invasive lingual excision (SMILE), as well as procedures that increase tension on the genioglossal and geniohyoid muscles (genial tubercle advancements and hyoid suspension). Unfortunately, consistency in the results was lacking and "risk to lingual artery and hypoglossal nerve prevented more aggressive tongue reduction techniques" [12].

Transoral robotic surgery (TORS) was first described by O'Malley et al. [13] in 2006 to treat upper aerodigestive tract neoplasms. In 2010, Vicini et al. [14] published their preliminary report on TORS for volumetric reduction of the tongue base in their multilevel surgical management strategy for treating OSA. This pilot clinical and cadaveric anatomical study led to widespread use of this new application.

The da Vinci robotic system (Intuitive Surgical, Sunnyvale, CA, USA), with the angled scope and EndoWrist® technology, allowed the surgeon to work around the angled corner in the small oropharyngeal cavity while maintaining agility and visibility of the surgical area, thus allowing precise excision of the obstructing tongue tissue. Here, we will review the current indications and contraindications, surgical anatomy of the lingual vasculature, surgical setup and techniques, and results and complications using TORS for tongue base reduction in the multilevel surgical treatment strategy for treating OSA.

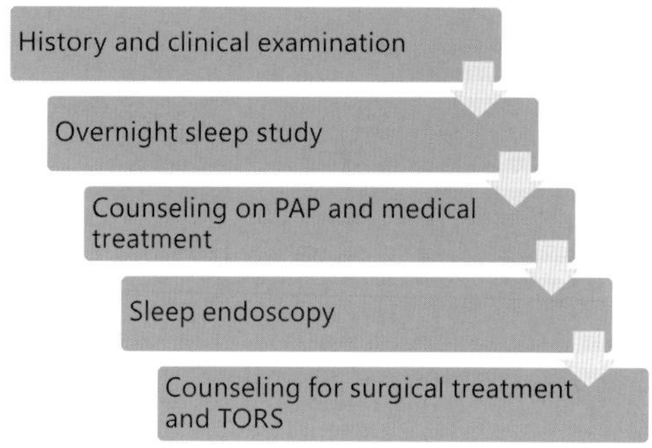

Fig. 1. Stepwise approach in the selection and counseling of patients for surgical treatment of obstructive sleep apnea and for TORS. PAP, positive airway pressure; TORS, transoral robotic surgery.

Assessment

Patients who have failed medical and PAP treatment for OSA with demonstrated tongue base obstruction during sleep endoscopy can be offered transoral robotic tongue base reduction. A stepwise approach (Fig. 1) starting with the patient history, clinical examination, sleep endoscopy, and counseling on options and results is required. The indications and contraindications for surgical treatment and TORS are listed in Table 1. Inadequate mouth opening and trismus is a contraindication for TORS.

Surgical Anatomy

Before conducting this surgical procedure, it is imperative to understand the anatomy of the lingual neurovascular bundle. The area of concern is the deep lingual artery, which is the terminal branch of the lingual artery. The average distance between the 2 deep lingual arteries in the substance of the tongue at the foramen caecum is 27.78 ± 6.57 mm, and the distance to the surface of the tongue is 29.27 ± 5.39 mm [15].

Table 1. Indications and contraindications for transoral robotic surgery

Indications	Contraindications
Base of tongue obstruction Lingual tonsillar enlargement Epiglottic collapse Supraglottic obstruction	Severe cardiopulmonary diseases Uncontrolled psychiatric diseases Unrealistic expectations Trismus Adequate mouth opening

Table 2. Equipment needed for transoral robotic surgery

da Vinci S/Si/Xi surgical system (surgeon console, patient cart, and vision cart) 30° robotic lens (12 mm, 8.5 mm) Maryland EndoWrist dissector (5 mm) Monopolar EndoWrist cautery (5 mm) Tonsillectomy set with Crow Davis mouth gag and Davis Meyer blade (assorted sizes) Laryngoforce vascular clip, applicator and clip Endolaryngeal microsurgery black sucker Silverglide bipolar forceps	Camera port, 5-mm cannula ×2 Antifog solution Tracheostomy set (standby) FK-WO TORS retractor set (standby)

The safe area for resection between the 2 deep lingual arteries at the tongue base is about 31 mm wide with the tongue in a resting position [16]. However, these distances change as the tongue is protruded out of the oral cavity and with compression with the tongue retraction blade. With the tongue pulled out in the extended/protruded position, the depth of the deep lingual artery on the tongue dorsum becomes less and the interarterial distance is narrowed compared to the resting position at the foramen caecum and 1 cm before and 1 cm after the foramen caecum. The deep lingual artery is brought closer to the tongue surface at these 3 points (from about 28 to 24–25 mm) and closer together (from about 20 to 14 mm) at the foramen caecum [17]. The TORS robotic surgeon should bear in mind the changes in the anatomy of the 2 lingual arteries and changes in the safety margin during surgery for adequate resection.

The dorsal lingual branches arising from the main lingual artery supply the mucosal of the dorsum of the tongue. These vessels generally do not cause troublesome bleeding and can be easily hemostased with bipolar diathermy or suction diathermy.

Surgical Equipment and Setup

The robotic and general equipment needed is listed in Table 2. The configuration of the surgical bed, robotic cart, surgeon's console, vision cart for the assisting surgeon to view the operation, and position of the assisting surgeon at the head of the table is shown in Figure 2. The da Vinci Robotic system is placed at about 30–45° to the long axis of the patient's bed. The anesthetic machine and anesthetist is on the left of the patient and should not obstruct the view of the assisting surgeon. The final position with the robotic arm, camera, and gag in situ is shown in Figure 2.

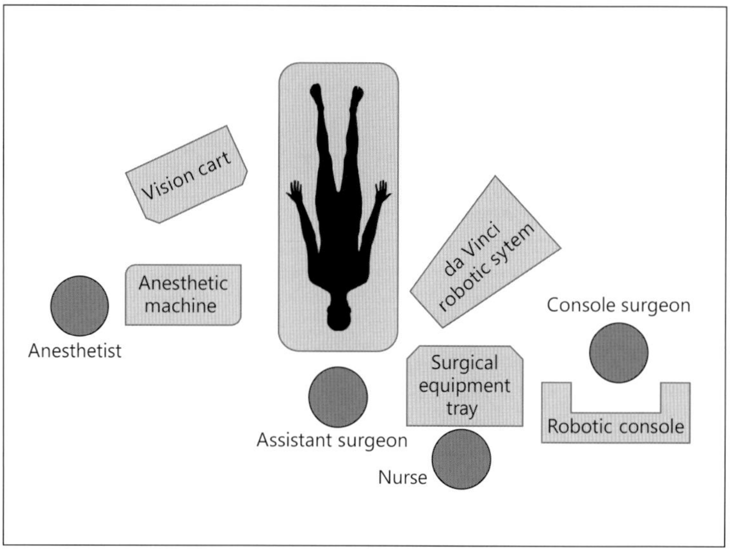

Fig. 2. Operating room setup for transoral robotic surgery using the da Vinci system.

Surgical Technique

Robotic tongue base reduction is performed as a component of multilevel surgery treatment in conjunction with other upper airway modification surgery (nasal, palatal, and pharyngeal) in the same setting. The technique, as first described by Vicini et al. [14], is outlined below.

The patient is intubated transnasally. Silk sutures are placed on the anterior third of the tongue to pull and bring forward the base of tongue for better visualization and access by the camera endoscope. The mouth gag is introduced into the oral cavity. The Davis Meyer blade and Crow Davis gag are sufficient for adequate exposure. The robotic system is moved into position, as shown in the setup. The 2 robotic arms are moved into position and a 5-mm cannula is fitted into the oral cavity, after which the 12 mm 30° up-facing endoscope is introduced. In patients with limited mouth opening, the 8.5 mm 30° up-facing endoscope may be needed. The da Vinci 5-mm EndoWrist® Maryland dissector and the monopolar diathermy are then introduced with the robotic arm.

Incision commences along the midline, starting from the foramen caecum, posterior inferior towards the vallecula. Once a trench of about 1 cm is created, retraction of the tissue can be achieved with Maryland forceps as the incision is further deepened. The plane of dissection and resection is created superficially to the muscles. The incision extends from the medial to lateral direction and caution is taken to avoid the laterally placed hypoglossal nerve and lingual artery. Depending on the extent of collapse and amount of tissue, a portion of muscle may be excised as well. The right and left hemibase of the tongue will be resected on either side. The volume of excised tissue can be measured via saline displacement.

A partial epiglottidectomy is performed if there is a posterior inspiratory collapse. Only the upper third of the epiglottis is removed to prevent aspiration.

An assisting surgeon will be seated at the head of the patient with access to the patient's oral cavity and oropharynx. He or she may need to help the console surgeon retract tissue by providing counter retraction for easier dissection and resection.

Postoperatively, the patient may be kept intubated for 1 day or so, or extubated in the operating room, and transferred to the high-dependency unit for observation. Patients are given intravenous steroids, antibiotics, and analgesia.

Outcome Measures

Transoral robotic tongue base reduction has become an established surgical technique in the multilevel surgical management of OSA hypopnea syndrome (OSAHS). Currently, there are many published articles with over 400 patients in various tertiary care centers all over the world that were treated for OSA with TORS using the da Vinci system.

Primary Outcomes

Three main outcome parameters have been considered. These are the AHI, lowest oxygen saturation (LSAT), and Epworth Sleepiness Scale (ESS).

AHI, LSAT, and ESS: Cure and Success

Based on the pooled results of 4 case series of palatal surgery and TORS for tongue base reduction and the analysis of weighted pooled data of 329 patients, the preoperative AHI decreased by 59.4% from 44.1 to 17.9 postoperatively [18–22]. In their study of 121 patients, Hoff et al. [23] reported surgical cure and success rates of 14 and 51.2%, respectively, and an AHI reduction from 42.7 to 22.2 following surgery.

Based on a definition of cure as reducing AHI to <5 and success as reducing AHI to <20 and a 50% reduction, the surgical cure and success for palatal surgery with TORS is 23.8 and 66.7%, respectively [18–22]. Pooled data analysis of the same 4-case series also showed that LSAT increased by 6.3%, from 79.0 to 84.0% [18–22]. The ESS was found to improve by 55.2%, from 12.8 to 5.7 [18–22].

Isolated Transoral Robotic Surgery for Tongue Base Reduction

The role of isolated TORS for tongue base reduction was investigated by Lin et al. [24]. In their case series of 12 subjects, they reported a statistically significant improvement in AHI of 56.2 ± 28.3%, from 43.9 ± 41.1 to 17.6 ± 16.2 ($p = 0.007$). The ESS also showed a statistical improvement of 53%, from 13.7 ± 5.2 to 6.7 ± 4.5 ($p < 0.001$). However, LSAT did not show a statistical improvement, increasing from 83.3 ± 5.5 to 84.0 ± 6.4% ($p = 0.680$). They included patients with tongue base obstruction who had other airway surgeries and primarily tongue base obstruction.

Prognostic Factors in Primary Outcome Measures

The preoperative BMI, AHI, the type of airway collapse, and the amount of tissue resected may influence the surgical outcomes of palatal surgery and TORS.

Influence of BMI on AHI, LSAT, and ESS

A reduction in BMI following upper airway surgery can confound the surgical outcome. This may be true for patients undergoing TORS, where reduced oral intake from dysphagia and odynophagia may cause a weight reduction.

Studies that looked at changes before and after TORS did not reveal statistical changes in BMI to account for the reduction in AHI. Hoff et al. [23] found that patients with a lower preoperative BMI had a higher chance of surgical success: the proportion of patients who had surgical success with BMIs of <30 and <25 were 56.5 and 78.3%, respectively. Lin et al. [22] found that patients with a BMI <30 respond better than those with a BMI >40. The surgical response rate of patients with a BMI <30 is 88.2% and those with a BMI >40 is 16.7%.

Preoperative AHI and Lateral Velopharyngeal Collapse

The severity of OSA seems to have an impact on the success and cure rate of palatal surgery and

TORS. Lin et al. [22] found that patients with a preoperative AHI <60 had a better response rate of 67.9% compared to those with an AHI >60, who had a response rate of only 18.2%. The absence of velopharyngeal collapse on sleep endoscopy seemed to be a good prognostic factor for surgical response. The same authors reported that patients who had lateral pharyngeal wall collapse had a response rate of 25.0%, whereas those without had a response rate of 66.7% [22]. This may be a useful criterion for patient selection and preoperative counseling. Chiffer et al. [25] used magnetic resonance imaging to evaluate how the airway and soft tissue volume change following surgery, discovering that changes in the retropalatal and total lateral wall soft tissue volume correlated significantly with an improvement in AHI.

Volume of Tissue Resected

The volume of tissue resection may correlate with surgical outcomes. Six case series in the literature studied the amount of tissue removed and correlated it with the surgical response. Friedman et al. [19] reported resecting an average of 2.28 g (SD 0.43) of tissue. They concluded that the weight of lingual tissue removed did not correlate with the improvement in AHI, LSAT, or ESS. Other studies, however, measured the resected volume in milliliters using the water displacement technique, and reported tissue resection of approximately 7–11 mL, with 1 study reporting resection of 22.2 mL. Eesa et al. [26] recommended that surgeons should resect 10–20 cc for optimal results and at least 7 mL to relieve tongue base obstruction. Hoff et al. [23] reported that a lower resection volume (6.07 mL) was significantly associated with a poorer outcome compared to a higher volume of resection (8.24 mL). Lin et al. [22] separately reported a positive correlation between the volume of base of tongue tissue resected and surgical success. Vicini et al. [27] reported that resection of 7 mL of tissue is required in order to observe a change. However, it is important to note that a larger patient may have more tissue and hence require a greater amount of tissue to be resected, and vice versa. Therefore, it is important not to follow these reported volumes of resection blindly.

Prior Airway Surgery

Thaler et al. [28] compared outcomes between patients who had failed previous surgery with those who had never undergone surgery. The total sleep time when oxygen saturation falls below 90% (hypoxic time) improved from 14 to 3.6% for patients who had no previous upper airway surgery versus the group that had previous surgery, for whom the hypoxic time improved from 21.1 to 12.5%. They concluded that the operated group was challenging to treat with poor outcomes. However, this must be interpreted with care as the greater hypoxic time may represent more severe disease and hence a greater normalizing difficult, rather than being a representation of previous surgery.

Secondary Outcomes

Secondary outcomes other than AHI and LSAT have been studied and need to be taken into account when interpreting the success of this treatment.

Polysomnographic Parameters

Toh et al. [21] and Thaler et al. [28] reported that polysomnographic parameters apart from AHI improved significantly following surgery. These parameters included the amount of N3 and rapid eye movement sleep, respiratory disturbance index, oxygen desaturation index, and the percentage duration of total sleep time with O_2 less than 90% (hypoxic time).

Subjective Outcomes Measures

A visual analog scale was used by some authors for the patient and bed partner to evaluate subjective outcomes, including intensity and satisfaction with the results of surgery. Friedman et al. [19] reported that the snoring visual analog scale

Table 3. Complications of transoral robotic surgery

Major	Minor
Bleeding	Bleeding
Dehydration	Tongue numbness
Oropharyngeal adhesions and stenosis	Dysgeusia
Dysphagia requiring prolonged intravenous fluids	Injury to tongue, teeth, gums, or temporal mandibular joint

decreased from a preoperative value of 9.1 ± 1.0 to 2.3 ± 3.9 postoperatively in their patients, while Toh et al. [21] also noted that the snoring intensity of their patients reduced similarly, from 8.7 ± 1.43 to 3.5 ± 1.7.

Robotic Time

The average robotic setup time reported in the literature was 24.8 min and average robotic operative time was 30.8 min [20–22]. With increasing experience, surgeons can expect a reduction in both the robotic setup and operative times [20, 21].

Length of Hospital Stay

The average length of stay was 3.2 days, ranging from 1.8 ± 1.3 to 9.5 ± 2.9 days. [18, 21, 24, 25, 29–32]. Glazer et al. [29] implemented home visits by a nurse following discharge. Patients with a poor oral intake and severe pain are readmitted to a short-stay unit for hydration and analgesic treatment. Patients are instructed to seek medical attention in the event of persistent pain and dehydration. However, Friedman et al. [19] reported that, although patients undergoing robotic surgery stayed an average of 0.5 nights longer compared to patients undergoing radiofrequency base of the tongue treatment, they did not experience more pain compared to the latter or compared to patients undergoing submucosal lingualplasty on postoperative day 1.

Complications

Complications can result from performing TORS, and can be classified as major if they warrant surgical intervention and/or readmission, and minor if they are treated and managed conservatively. No perioperative death associated with the use of the robotic system has been reported. Complications, according to the published literature, are tabulated in Table 3.

Tracheostomy

The main concern in TORS is the risk of injury to the lingual artery causing massive intra- or postoperative bleeding. This can be potentially difficult to stop because of the anatomical constraint and, hence, can compromise the airway. Excessive edema may also cause upper airway obstruction. Vicini et al. [14, 27] routinely performed tracheostomy on all of their TORS patients. Retrospectively, they concluded that this might not be necessary. In some centers, sleep surgeons continued to perform elective tracheostomy whilst others kept the patient intubated overnight, while some surgeons extubate patients in the operating room after the surgery but keep them monitored in the surgical intensive care unit or high-dependency unit [14, 18–24]. Emergency tracheostomy was never required in the published studies.

Major Bleeding and Self-Limiting Bleeding

Bleeding due to injury of the lingual artery can be life threatening. Vicini et al. [20] reported that the risk of intraoperative bleeding is 0.4%. Based on published results, the volume of blood lost intraoperatively ranged from 12.9 to 27.7 mL. Friedman et al. [19] described the use of preoperative Doppler ultrasound to delineate the position of the lingual artery to prevent injury. This was required because they used the triangular mucosal sparing resection technique, which potentially posed a higher risk for the lingual vasculature. This additional step is not routinely necessary when the traditional inverted pyramid technique is used.

Postoperative bleeding can be classified as major or minor, depending on whether surgical intervention under anesthesia was needed. Using this definition, Vicini et al. [20] estimated that the rates of major and minor postoperative bleeding were 0 and 15%, respectively. In published studies, the average rates are 2.9% for major bleeding and 2.0% for minor bleeding [18]. Major bleeding may require surgical intervention. An angled endoscope (usually 30°) is needed to visualize the tongue base area adequately following intubation and a regular setup similar to that for TORS. Suction diathermy or bipolar diathermy is usually needed to stop the bleeding, and vascular clips may be needed.

Bleeding may result from the tonsillectomy site rather than the tongue base [21]. It may be associated with an early restart of anticoagulants the patient uses because of a concomitant comorbidity [29]. Self-limiting bleeding can occur and usually observation and conservative management is sufficient.

Oropharyngeal Stenosis

Oropharyngeal stenosis causing oropharyngeal stenosis has been described by some authors, albeit rarely [18, 20, 33]. These patients required revision surgery, scar lysis, and triamcinolone injection. Excessive cautery use and dissection of the tonsillar inferior pole, and communicating the wound with the tongue base where the lingual tonsils join with the tonsils may be the causes.

Pain and Dehydration

Dehydration secondary to a poor oral intake as a result of pain is a common complication postoperatively. About 5–10% of patients may require intravenous fluids [28, 29]. In the case series reporting patients needing intravenous fluids, none required nasogastric tube feeding.

Pain can be adequately controlled with acetaminophen and nonsteroidal anti-inflammatory analgesia, and in some centers a judicial use of opioids. Some authors reported the use of gabapentin to modulate pain [23, 28, 29]. Toh et al. [21] reported that the mean first postoperative day on which pain was measured by visual analog scale was 6.5 and that this was usually sufficiently controlled with oral paracetamol and nonsteroidal anti-inflammatory drugs. The majority of patients can have an oral intake of food by postoperative day 2.

Dysphagia

Swallowing may be compromised after TORS because of the resection of tongue tissue, the lingual tonsils, and/or the epiglottis. Patients are generally able to gradually tolerate oral feeding on postoperative day 1 and their diet of choice within 2 weeks after surgery [20, 21, 26]. No case report of hypoglossal nerve injury has been reported in the literature.

The amount of cautery use may be related to the presence of dysphagia and is generally self-limiting [19, 29]. An objective study of swallowing function using gastrografin showed the presence of aspiration in 24.4% of patients, but this resolved fully after 3 months [27]. The same researchers administered the MD Anderson Dysphagia Inventory and found no significant difference (p = 0.56) in self-reported swallowing function pre- and postoperatively. The aspiration has no relation to the amount of tissue resection and does not result in pneumonia or any significant weight loss. The problem of dysphagia is generally self-limiting and resolves within 3 months [27].

Dysgeusia and Tongue Numbness

The most common minor complication was dysgeusia/altered sense of taste and tongue numbness. The incidence is between 14.2 and 100% [20, 21, 26]. Possible mechanisms could be direct injury to the taste buds in the anterior tongue, resection of circumvallate papilla at the base of the tongue, prolonged retraction causing stretching and neuropraxia of the lingual nerve or glossopharyngeal nerve, and prolonged pressure from the tongue blade [20, 21]. Periodic interval relaxation and reducing the total robotic operating time may be necessary to minimize this complication, which is generally transient in nature and resolves within 3 months to 1 year.

Other Complications

Other complications reported in the literature that may be related to the underlying comorbid status or from the intubation process and perioperative anesthetic care include cardiac arrhythmias, hypoxemia, pulmonary embolism, reintubation, pneumonia, gum injury, and pharyngeal lacerations [30–32, 34].

Comparing Transoral Robotic Surgery to Other Tongue Base/Hypopharyngeal Procedures

Tongue base procedures or hypopharyngeal surgeries can be classified into volumetric tissue reduction procedures (radiofrequency, coblation, SMILE, midline glossectomy) and tissue tensioning procedures (genioglossal advancement, hyoid suspension, hyoid expansion). Comparing surgical outcomes between TORS and other hypopharyngeal surgeries is complicated because different tongue base procedures are devised to treat different aspects of upper airway collapse, while tissue volumetric reduction and tissue tensioning surgical procedures need not be mutually exclusive. Maxillomandibular advancement is a global airway treatment strategy and cannot be fairly compared to TORS for tongue reduction.

Kezirian and Goldberg [12] assessed all the various hypopharyngeal surgical outcomes and reported unpredictable results with varied success rates, ranging from as low as 20% to as high as 83%. Murphey et al. [35] performed a systematic review and meta-analysis of glossectomy for OSAHS, which included other minimally invasive procedures and TORS. They reported that the AHI reduced from 48.1 ± 22.01 preoperatively to 19.05 ± 15.46 postoperatively.

Friedman et al. [19] compared the outcomes of TORs for tongue base reduction with patients who had radiofrequency of the tongue base (RFBOT) and with patients who underwent SMILE. All 3 groups had Z-palatopharyngoplasty. All patients reported an improvement in AHI and LSAT. The AHI reduction for the TORS group was 60.5 ± 24.9 versus 37.0 ± 51.6% (p = 0.042), and 32.0 ± 43.3% (p = 0.012) for SMILE and RFBOT, respectively. However, oxygen saturation improvement was observed only in the robotic group. The surgical success rate for TORS (66.7%) was statistically significant compared with RFBOT (20.8%, p = 0.001). When compared to the success rate of SMILE (45.5%, p = 0.135), the result was not statistically significant although the TORS group achieved a higher percentage. The time taken to resume a normal diet and activity appears to be longer in the robotic group compared to the other groups.

To really compare and study the difference between tongue base tissue volumetric surgical procedures, a matched randomized controlled trial is necessary, thus providing an impetus for future research. In patients with mandibular skeletal restriction and/or lateral obstruction of the hypopharyngeal muscles, genial tubercle surgical advancements (mortised genioplasty and genial tubercle advancement procedures) and hyoid suspension/expansion (to the mandible or thyroid cartilage) change the skeletal framework and work by altering muscle tension. Comparing tongue volumetric reduction with tongue and hypopharyngeal tissue tensioning is unsound because of the fundamental difference in the way they work in their utilization. They should be considered complementary and not mutually exclusive.

Future Work and Advances

The single-port da Vinci robotic system (Intuitive Surgical, Sunnyvale, CA, USA; http://www.intuitivesurgical.com/) may be able to overcome the current limitations of patients with a small mouth opening and trismus. With the endoscope and working arms coming from the same port, this new innovation may provide better ease of access to the narrow oral cavity and tongue base in OSA patients.

The new Flex® robotic system from Medrobotics (Raynham, MA, USA) recently obtained FDA approval for use during transoral procedures [36]. This minimally invasive system also enables surgical access and visualization in difficult to assess sites through a single port. The system uses a flexible robotic scope that is able to move through twists and turns in the human oral cavity and pharynx. The high-definition vision system and flexible instruments then allow the surgeon to operate in confined areas.

Conclusion

Technological advancement has allowed tongue base reduction causing airway obstruction to be addressed using the da Vinci robotic system safely and with ease. It is merely a tool to allow resection of the tongue base under direct vision and allows the surgeon to work around corners in the oral cavity and hypopharynx. Safe resection of an obstructing tongue base achieves improved AHI, ESS, and LSAT, with acceptable morbidity. An appropriate patient selection is paramount in yielding desirable results.

References

1 Garvey JF, Pengo MF, Drakatos P, Kent BD: Epidemiological aspects of obstructive sleep apnea. J Thorac Dis 2015;7: 920–929.
2 Gami AS, Olson EJ, Shen WK, et al: Obstructive sleep apnea and the risk of sudden cardiac death: a longitudinal study of 10,701 adults. J Am Coll Cardiol 2013;62:10.
3 Kendzerska T, Mollayeva T, Gershon AS, et al: Untreated obstructive sleep apnea and the risk for serious long-term adverse outcomes: a systematic review. Sleep Med Rev 2014;18:49–59.
4 Lee W, Nagubadi S, Kryger MH, Mokhlesi B: Epidemiology of obstructive sleep apnea: a population-based perspective. Expert Rev Respir Med 2008;2: 349–364.
5 Weaver TE, Grunstein RR: Adherence to continuous positive airway pressure therapy: the challenge to effective treatment. Proc Am Thorac Soc 2008;5:173–178.
6 Campos-Rodriguez F, Peña-Griñan N, Reyes-Nuñez N, De la Cruz-Moron I, Perez-Ronchel J, De la Vega-Gallardo F, Fernandez-Palacin A: Mortality in obstructive sleep apnea-hypopnea patients treated with positive airway pressure. Chest 2005;128:624–633.
7 Aurora RN, Casey KR, Kristo D, Auerbach S, Bista SR, Chowdhuri S, Karippot A, Lamm C, Ramar K, Zak R, Morgenthaler TI; American Academy of Sleep Medicine: Practice parameters for the surgical modifications of the upper airway for obstructive sleep apnea in adults. Sleep 2010;33:1408–1413.
8 Caples SM, Rowley JA, Prinsell JR, Pallanch JF, Elamin MB, Katz SG, Harwick JD: Surgical modifications of the upper airway for obstructive sleep apnea in adults: a systematic review and meta-analysis. Sleep 2010;33:1396–1407.
9 Lin HC, Friedman M, Chang HW, Gurpinar B: The efficacy of multilevel surgery of the upper airway in adults with obstructive sleep apnea/hypopnea syndrome. Laryngoscope 2008;118:902–908.
10 Ravesloot MJ, de Vries N: One hundred consecutive patients undergoing drug-induced sleep endoscopy: results and evaluation. Laryngoscope 2011;121: 2710–2716.

11 Vroegop AV, Vanderveken OM, Boudewyns AN, Scholman J, Saldien V, Wouters K, Braem MJ, van de Heyning PH, Hamans E: Drug-induced sleep endoscopy in sleep-disordered breathing: report on 1,249 cases. Laryngoscope 2014;124:797–802.
12 Kezirian EJ, Goldberg AN: Hypopharyngeal surgery in obstructive sleep apnea: an evidence-based medicine review. Arch Otolaryngol Head Neck Surg 2006; 132:206–213. Review
13 O'Malley BW Jr, Weinstein GS, Snyder W, Hockstein NG: Trans-oral robotic surgery (TORS) for base of tongue neoplasms. Laryngoscope 2006;116:1465–1472.
14 Vicini C, Dallan I, Canzi P, Frassineti S, La Pietra MG, Montevecchi F: Transoral robotic tongue base resection in obstructive sleep apnoea-hypopnoea syndrome: a preliminary report. ORL J Otorhinolaryngol Relat Spec 2010;72:22–27.
15 Li SH, Shi HJ: Lingual artery CTA-guided midline partial glossectomy for treatment of obstructive sleep apnea hypopnea syndrome. Acta Otolaryngol 2013; 133:749–754.
16 Hou T, Shao J, Fang S: Laryngoscope – the definition of the V zone for the safety space of functional surgery of the tongue. Laryngoscope 2012;122:66–70.
17 Wu DH, Qin J, Guo XH, Li SH: Analysis of the difference in the course of the lingual arteries caused by tongue position change. Laryngoscope 2015;125: 762–766.
18 Rangabashyam M, Huang WJ, Hao Y, Han HJ, Loh S, Toh ST: State of the art transoral robotic surgery for obstructive sleep apnea. Robotic Surg Res Rev 2016; 3:13–28.
19 Friedman M, Hamilton C, Samuelson CG, Kelley K, Taylor D, Pearson-Chauhan K, Maley A, Taylor R, Venkatesan TK: Transoral robotic glossectomy for the treatment of obstructive sleep apnea-hypopnea syndrome. Otolaryngol Head Neck Surg 2012;146:854–862.
20 Vicini C, Montevecchi F, Campanini A, Dallan I, Hoff PT, Spector ME, Thaler E, Ahn J, Baptista P, Remacle M, Lawson G, Benazzo M, Canzi P: Clinical outcomes and complications associated with TORS for OSAHS: a benchmark for evaluating an emerging surgical technology in a targeted application for benign disease. ORL J Otorhinolaryngol Relat Spec 2014;76:63–69.
21 Toh ST, Han HJ, Tay HN, Kiong KL: Transoral robotic surgery for obstructive sleep apnea in Asian patients: a Singapore sleep centre experience. JAMA Otolaryngol Head Neck Surg 2014;140: 624–629.
22 Lin HS, Rowley JA, Folbe AJ, Yoo GH, Badr MS, Chen W: Transoral robotic surgery for treatment of obstructive sleep apnea: factors predicting surgical response. Laryngoscope 2015;125:1013–1020.
23 Hoff PT, Glazer TA, Spector ME: Body mass index predicts success in patients undergoing transoral robotic surgery for obstructive sleep apnea. ORL J Otorhinolaryngol Relat Spec 2014;76:266–272.
24 Lin HS, Rowley JA, Badr MS, Folbe AJ, Yoo GH, Victor L, Mathog RH, Chen W: Transoral robotic surgery for treatment of obstructive sleep apnea-hypopnea syndrome. Laryngoscope 2013;123: 1811–1816.
25 Chiffer RC, Schwab RJ, Keenan BT, Borek RC, Thaler ER: Volumetric MRI analysis pre- and post-transoral robotic surgery for obstructive sleep apnea. Laryngoscope 2015;125:1988–1995.
26 Eesa M, Montevecchi F, Hendawy E, D'Agostino G, Meccariello G, Vicini C: Swallowing outcome after TORS for sleep apnea: short- and long-term evaluation. Eur Arch Otorhinolaryngol 2015; 272:1537–1541.
27 Vicini C, Dallan I, Canzi P, Frassineti S, Nacci A, Seccia V, Panicucci E, Grazia La Pietra M, Montevecchi F, Tschabitscher M: Transoral robotic surgery of the tongue base in obstructive sleep apnea-hypopnea syndrome: anatomic considerations and clinical experience. Head Neck 2012;34:15–22.
28 Thaler ER, Rassekh CH, Lee JM, Weinstein GS, O'Malley BW Jr: Outcomes for multilevel surgery for sleep apnea: obstructive sleep apnea, transoral robotic surgery, and uvulopalatopharyngoplasty. Laryngoscope 2016;126:266–269.
29 Glazer TA, Hoff PT, Spector ME: Transoral robotic surgery for obstructive sleep apnea: perioperative management and post-operative complications. JAMA Otolaryngol Head Neck Surg 2014;140:1207–1212.
30 Vicini C, Montevecchi F, Magnuson JS: Robotic surgery for obstructive sleep apnea. Curr Otorhinolaryngol Rep 2013; 1:130–136.
31 Muderris T, Sevil E, Bercin S, Gul F, Kiris M: Transoral robotic lingual tonsillectomy in adults: preliminary results. Acta Otolaryngol 2015;135:64–69.
32 Lee JM, Weinstein GS, O'Malley BW Jr, Thaler ER: Transoral robot-assisted lingual tonsillectomy and uvulopalatopharyngoplasty for obstructive sleep apnea. Ann Otol Rhinol Laryngol 2012;121: 635–639.
33 Muderris T, Sevil E, Bercin S, Gul F, Kiris M: Oropharyngeal stenosis after transoral robotic lingual tonsillectomy. J Craniofac Surg 2015;26:853–855.
34 Hoff PT, D'Agostino MA, Thaler ER: Transoral robotic surgery in benign diseases including obstructive sleep apnea: safety and feasibility. Laryngoscope 2015;125:1249–1253.
35 Murphey AW, Kandl JA, Nguyen SA, Weber AC, Gillespie MB: The effect of glossectomy for obstructive sleep apnea: a systematic review and meta-analysis. Otolaryngol Head Neck Surg 2015;153: 334–342.
36 Remacle M, Prasad VM, Lawson G, Plisson L, Bachy V, Van der Vorst S: Transoral robotic surgery (TORS) with the Medrobotics Flex™ System: first surgical application on humans. Eur Arch Otorhinolaryngol 2015;272:1451–1455.

Song Tar Toh, MBBS, MRCSEd, MMED (ORL), MMED (Sleep Medicine), FAMS (ORL)
Sleep Apnea Surgery Service, Department of Otolaryngology, Singhealth Duke-NUS Sleep Centre
Singapore General Hospital, Outram Road
Singapore 169608 (Singapore)
Email : toh.song.tar@singhealth.com.sg

Lin H-C (ed): Sleep-Related Breathing Disorders. Adv Otorhinolaryngol. Basel, Karger, 2017, vol 80, pp 136–144
DOI: 10.1159/000470885

Pediatric Obstructive Sleep Apnea: Where Do We Stand?

Yu-Shu Huang[a, b] • Christian Guilleminault[c]

[a]Department of Child Psychiatry, Chang Gung Memorial Hospital and College of Medicine, and [b]Craniofacial Research Center and Sleep Center, Chang Gung Memorial Hospital and University, Taoyuan, Taiwan; [c]Stanford University Sleep Medicine Division, Redwood City, CA, USA

Abstract

Pediatric obstructive sleep apnea (OSA) was initially described in 1976. In 1981, Dr. Guilleminault emphasized that pediatric OSA was different from the clinical presentation reported in adults. It was characterized by more disturbed nocturnal sleep than excessive daytime sleepiness, and presented more behavioral problems, particularly school problems, hyperactivity, nocturnal enuresis, sleep terrors, depression, insomnia, and psychiatric problems. The underlying causes of pediatric OSA are complex. Such factors as adenotonsillar hypertrophy, obesity, anatomical and neuromuscular factors, and hypotonic neuromuscular disease are also involved. Adenotonsillectomy (T&A) has been the recommended treatment for pediatric OSA, but in the recent past this practice has been placed very much in question. Therefore, we will discuss the mechanism of pediatric OSA and investigate obese and nonobese pediatric sleep-disordered breathing. Moreover, the important concept that dysfunction leads to the dysmorphism that impacts on the size of the upper airway has been advanced recently. Finally, the treatments of pediatric OSA, such as T&A, medication, the orthodontic approaches (rapid maxillary expansion, or mandibular advancement with functional appliances), positive airway pressure, and noninvasive treatment, such as myofunctional therapy (MFT), will be investigated. A "passive MFT" has been tried recently, but very few results exist. In conclusion, we have made progress in our understanding of pediatric OSA, and we can even recognize factors leading to its development or worsening. However, pediatricians and pediatric subspecialists are often unaware of the advances and the remedies available.

Pediatric Sleep-Disordered Breathing

Pediatric obstructive sleep apnea (OSA) was initially described in 1976 [1]. In 1981, Guilleminault et al. [2] published a review of 50 pediatric patients and emphasized that pediatric OSA was different from the clinical presentation reported in adults. The authors emphasized that these children had more disturbed nocturnal sleep than excessive daytime sleepiness, and presented more behavioral problems, particularly school problems related to attention deficit, poor school performance, hyperactivity, symptoms classified as "attention-deficit-hyperactivity syndrome," nocturnal enuresis, sleep terrors, sleepwalking and confusional arousals, symptoms classified as "NREM parasomnias," depression, insomnia, and psychiatric problems. Cardiology-related symptoms were infrequent, but tachybradycardia was regularly noted. Adenotonsillectomy (T&A) was performed and successful in some but not all children, as was clearly demonstrated on follow-up. A small group of children presented an abnormal weight increase after T&A. These children presented apnea and hypopneas closely following the current polysomnographic definition. However, a year later Guilleminault et al. [3] published a new report indicating that children may present the same chronic symptoms, yet polysomnographic investigations performed with these children using esophageal pressure manometry showed an absence of apnea and hypopnea, but the presence of abnormal upper airway (UA) resistance and more or less snoring. In 1982, many of the features presented today in reports on pediatric sleep-disordered breathing (SDB) were already clearly indicated and some of the issues still need further research. These include recurrence post-T&A, weight increase following T&A, and the issue of having SDB with similar complaints, symptoms, and clinical findings at evaluation associated with and without snoring, and with very different patterns of abnormal breathing at the polysomnography evaluation.

The obesity epidemic became relevant in the 1990s, adding a further level of complexity. Two different syndromes were observed in the same individual: obesity per se could lead to the same complaints and symptoms as OSA syndrome in a normal-weight child, and obesity could lead to the development of OSA as a comorbidity due to the deposit of fat in the tongue tissues and other UA muscles. This then could lead to a chest-bellow syndrome related to the abdominal fat deposit, and it could worsen the symptoms seen in a slim OSA child. Attributing the respective responsibilities to obesity and OSA in their clinical presentation was difficult, particularly due to the fact that often the child is not seen early, but only after several years of evolution.

Obesity and Sleep-Disordered Breathing

Obesity is a complex disorder leading to worsening supine ventilation secondary to restrictive chest-bellows syndrome [4]. Obesity also leads to progressive fatty infiltration of the neck and UA. MRI studies have shown that a progressive fatty infiltration of the geniohyoid and genioglossus muscles occurs along with the dissociation of muscle fibers with fat cells [5]. Certain ethnicities, particularly African-American children, have a stronger association between obesity and SDB [6].

Obesity is associated with a progressive dysfunction of the adipocytes. Preadipocytes differentiate into mature adipocytes and form adipose tissue in response to a positive energy balance. Adipose tissue not only stores energy, but also acts as a dynamic endocrine organ, vital for hormone and cytokine (adipokine) secretion. White adipose tissue, located in abdominal and subcutaneous deposits in mammals, performs the majority of energy storage and adipokine secretion [6]. Brown adipose tissue mediates the nonshivering thermogenesis, well known to protect infants from cold exposure. Genetics play a role in the control and development of white adipose tissue and brown adipose tissue. Dysfunction of adipocytes leads to the stimulation of adipokines, particularly TNF-α and interleukins 6 and 1.

These defects lead to pivotal inflammatory responses, both local and general, in addition to abnormal secretion of peptides found not only in the adipocyte, but also in the gut and brain. Peptides such as leptin, adiponectin, obesin, etc., are involved, and dysfunction of the adipocytes leads to leptin resistance and ghrelin dysfunction. These 2 peptides are crucial to food intake, insulin resistance, and the dysregulation of glucose and lipid control [7]. Overweight and obese individuals, with or without SDB, will develop these dysfunctions.

The consequences of these abnormalities affect the cardiovascular, respiratory, metabolic, and cerebral systems. Sleep fragmentation, which occurs with abnormal breathing, will cause changes in metabolic controls in part through the process of epigenetics, by which environmental events trigger a genetic cascade that would not have otherwise occurred. Obesity along with fatty infiltration of the UA will always lead to SDB from simple flow limitation to frank OSA.

Why Does the Upper Airway Collapse during Sleep in Nonoverweight Children?

Pharynx and Internal Factors

The pharynx is a collapsible tube: unlike the lower airways, there is no rigid support. The UA, consisting of skeletal muscles and soft tissues, supports nonrespiratory functions, such as sucking, swallowing, and vocalization/phonation, etc. Sleep causes fundamental modifications of pharyngeal muscle tone and reflex responses and can lead to narrowing and increased UA resistance in normal individuals.

The control of muscle tone during wakefulness and during sleep is different; this is true for the muscles constituting the walls of the UA in humans. There are 2 sleep states: rapid eye movement (REM) sleep and non-REM sleep; muscle control differs during the 3 states of alertness: wakefulness, and non-REM and REM sleep, respectively. In physiological terms, REM sleep is associated with the greatest amount of inhibition of volitional muscle tone. Sleep favors UA collapse, particularly due to the loss of tonic activation of UA muscles at the end expiration. Also, sleep usually occurs in a recumbent position and the degree of recumbence has an impact on the size of the UA, with lying flat on one's back being the position leading to the largest amount of change, compared to an erect position, due to the action of gravity and atmospheric pressure. The UA presents an intrinsic collapsibility that can be modelled as a "collapsible tube," with maximum flow (Vmax) determined by upstream nasal pressure (Pn) and resistance (Rn); the tube collapses and airflow stops at the critical pressure (Pcrit).

Pharynx and External Factors

External factors impact on the size of the UA, particularly when it is in a retropalatal or retroglossal position. Three of these factors are particularly prominent: fat deposits (related to the body mass index), craniofacial features (related to genetic and functional factors), and hypertrophied tissues, in part related to local inflammation. These external factors can be influenced by genetic and environmental factors.

Bone structure has an important role in the size of the UA. The development of the face is a very closely regulated event, with continuous interaction between the development of the entire brain, the skull, and the skull base. The growth of the transversal portion of the nasomaxillary complex is influenced by 3 factors: the development of the nasal fossae during fetal life, the growth of the ocular cavities related to ocular development during fetal life, and the activity of the intermaxillary suture that utilizes an endochondral mode of ossification and is active until about 16 years of age, and then undergoes complete synostosis by age 25 years. The face is located at the anterior-most point of the skull base and is therefore especially dependent on the processes involved in its growth, with the maxilla and mandible being "pushed forward" by the development of the skull base. The interaction between the developments of the nasomaxillary complex and the support of the head in an individual with vertical posture is a key adjustment.

Genetic factors are critical in this development. Most of the growth of the skull base is cartilaginous growth, and growth occurs in relation to "synchondroses" [8]. These serve as the sites of bone growth in the skull base and are located in the sutures between the bones forming the skull and skull base. Sphenoidal chondrosis is responsible for the vertical growth of the skull base. The skull base has an oblique direction and lowers the location of the occipital lobe, thereby affecting facial growth. The growth of the nasomaxillary complex is related not only to the sphenoido-occipital synchondroses, but also to the activity of the synchondroses of the skull base, and particularly the cleft at the following sutures: intermalar, intermaxillary, interpalatine, maxillomalar, and temporomalar.

Postnatal Activity

It is important to note that the intermaxillary suture is active postnatally, as mentioned above, and is influenced by specific functions, such as suction, mastication, swallowing, and nasal breathing. These functions mobilize the facial muscles that play a clear role in facial growth. The development of these functions is influenced by the quality of nasal respiratory roles, dental development, which involves the position and height of the alveoli and teeth position, and the activity and strength of the tongue and facial muscles. The vertical growth of the nasomaxillary complex is related to the activity of the posterior skull base, and also to that of the frontomalar, frontomaxillary, and maxillomalar sutures. It is also related to the position of the hard palate and alveolodental activity [8]. While the mandible is involved in the space controlling the size of the UA, it is independent of the base of the skull and is instead associated with the cervicothoracic-digestive axis. This structure involves many muscle and ligament attachments and dictates the head posture.

The Role of 2 Synchondroses Active Postnatally: The Intermaxillary and Alveolodental Synchondroses

Intermaxillary Synchondroses

The recognition of genetic impairments of endochondral growth leading to SDB is often delayed until after childhood. Ehlers-Danlos syndrome [9] is secondary to either an autosomal-dominant, autosomal-recessive, or X-linked mutation of genes located on proteins or enzymes, most commonly COL-1A1, COL 5A1, or 5A2. Clinical evaluation demonstrated the presence of an abnormally long face, narrow and high hard palate, and frequently associated crossbite. While initially only abnormalities of the nasomaxillary complex may be seen, as patients enter adulthood and develop worsening SDB, defects of the condyle may also be detected.

Alveolodental Synchondroses

When permanent teeth are absent or are extracted in early life during their growth period, this can lead to bone retraction and affect facial bone growth. There is an association between teeth agenesis and the presence of OSA in nonsyndromic children. Dental agenesis is linked to genetic mutations, with a dental homeo-code for the agenesis of canine, incisor, and molar teeth. The association between congenitally missing teeth and facial skeletal changes with "a straight to concave profile, pointed chin, reduced lower facial height and altered dental inclination" was noted by Ben-Bassat and Brin [10]. This was confirmed by more recent studies [11]. Dental research has shown that tooth agenesis is a common congenital disorder; it may be associated with syndromes, but it is also often seen in nonsyndromic children and its prevalence has varied, depending on the author, with findings oscillating between 10 and 20% of the studied populations [12]. In 10% of agenetic cases, 2 teeth are involved, with the 2nd premolar and the lateral incisor being considered as more frequent cases of agenesis, and 1–2% having oligodontia. There is an important role for abnormal craniofacial growth in the development of pediatric SDB and involvement of synchondroses, particularly in those still active during childhood.

Craniofacial Muscle Activity, Genes, and Abnormal Orofacial Growth

There is an interaction between muscle activities, particularly those of the face, and the growth and normal development of the UA. Genetic abnormalities impairing the normal activity of the striatal muscles, including facial muscles, lead to SDB. The most studied genetic disorder involving mutations and generalized muscle impairment is myotonic dystrophy, both type I and type II [13].

The results of an environmental impairment of orofacial muscle activity experiment involving monkeys [14, 15] suggested that nongenetic postnatal impairment may have an impact similar to genetically induced muscle impairment. The experimental data showed the presence of a continuous interaction between abnormal nasal resistance and orofacial growth through the intermediary of abnormal muscle tone and mouth breathing (with a change in the mandibular condyle position). The abnormal growth leads to further worsening of the nasal resistance. The consequence is a small UA.

There is a syndrome, known to be familial (although its genetic origin has not been demonstrated to date), which is clearly associated with the development of a small UA and SDB: the short lingual frenulum syndrome (ankyloglossia) [16, 17]. A short lingual frenulum has been associated with difficulties in sucking, swallowing, and speech. However, the oral dysfunction induced by a short lingual frenulum can lead to oral-facial dysmorphism, which decreases the size of UA support. The lingual frenulum is a vestigial embryological element that is mostly fibrous in its consistency as a result of adhesion between the tongue and the floor of the mouth during embryogenesis. Apoptosis controlled by genes separates the tongue from the primitive pharynx during embryogenesis [16–19]. "Clipping" of the short lingual frenulum is still proposed when difficulties are recognized early in life, but long-term results are reported as unpredictable when "clipping" is performed after the first few months of life. A short lingual frenulum modifies the position of the tongue. The orthodontic impact of this abnormal position may result in an anterior and posterior crossbite, a disproportionate growth of the mandible, and an abnormal growth of the maxilla [20, 21]. The tongue is normally placed high in the palate, and the continuous activity related to sucking, swallowing, and masticating induces stimulation of intermaxillary synchondrosis, as already mentioned. The interaction between abnormal bone growth stimulation and an absence of nasal breathing with secondary development of mouth breathing is responsible for the abnormal development of the oral-facial bone structures supporting the UA, thus increasing the risk of UA collapse during sleep. The abnormal oral-facial growth leading to a reduction in the ideal size of the UA occurs at a variable speed depending on the individual, and abnormal breathing during sleep occurs over time, with initial flow limitation, then progressive worsening toward full-blown OSA syndrome.

Nongenetic Impairment of Muscles and Abnormal Oral-Facial Growth

In children, prematurity is often associated with generalized muscle hypotonia. Its severity is dependent on the degree of prematurity, in spite of the disappearance of the diaphragmatic apneas of prematurity [22]. The development of OSA is observed. This atypical breathing pattern is associated with the development of mouth breathing and a high and narrow hard palate. Early premature infants often have abnormalities involving feeding functions, such as suction, mastication, and swallowing, with weakness of orofacial muscles that negatively alter craniofacial growth and lead to a small UA.

Functional Dysfunctions

These different studies demonstrate that impairment of the growth of the oral cavity and oral-facial structures early in life leads to the development of an abnormal anatomy of the bone support of the UA, increasing the risk of UA collapse during sleep. Some specific dysfunctions involving muscle tone lead to abnormal functions that impact on bone development of the structures supporting the UA; these functions include sucking, masticating, swallowing, and nasal breathing. Abnormal nasal breathing leads to mouth breathing, which is another dysfunction.

The concept that dysfunction leads to the dysmorphism that impacts on the size of the UA has recently been advanced. This concept has led to different treatment initiatives with the goal of: (a) demonstrating the negative effect of not addressing the dysfunction when treating OSA, and (b) trying to address the dysfunction directly and as early as possible.

Treatments and Outcomes

Negative Effect of Not Addressing the Dysfunction

For years, T&A has been the recommended treatment for pediatric OSA, but in the recent past this practice has been placed very much in question. First, many studies have shown that the use of T&A in pediatric OSA patients may have variable results, reaching an AHI of 1 or less in about 50% of cases (and as low as 32% in obese children) [23–27]. However, a long-term follow-up study [28] performed first in 6- to 12-year-old children with OSA and repeated in 4- to 6-year-old children with OSA who underwent T&A showed progressive recurrence and worsening in both "apparently cured" and "significantly improved" children, and in 68% of the children after 36 months of follow-up in the first study. Recurrence was also noted in the second study, except that progressive worsening was slower in the younger children (4- to 6-year-old group). The issue of the role of T&A has been raised by others. The CHAT study looking at children with low but abnormal AHI showed that a delay in performing T&A may not show the same polysomnography results as in the initial investigation. Recent studies of children post-T&A have shown that despite T&A removal, children may still present mouth breathing [11]. This mouth breathing may be related to the fact that children who have been mouth breathers for a certain time due to obstruction of the nose and UA have a "disuse" of their nose when breathing, and removal of obstructive UA tissues does not mean a systematic return to normal nasal breathing during sleep [11]. This mouth breathing may also be related to the progressive occurrence of dysmorphism secondary to the presence of mouth breathing that impacts on maxillary and mandibular growth, as shown in the experiment preformed on Rhesus monkeys [14, 15], or to dysmorphism related to congenital dental agenesis [29, 30] or other secondary dysfunctions (such as a short lingual frenulum or prematurity, etc.).

Addressing the Dysfunction as Early as Possible

The introduction of orthodontic treatment (rapid maxillary expansion or bimaxillary expansion) [31, 32] has also shown that some children may not need T&A. The orthodontic treatment of dysmorphism that leads to a small UA (again with rapid maxillary expansion or other orthodontic approaches) may be sufficient to avoid T&A and restore nasal breathing during sleep [31, 32]. However, both orthodontics and T&A may be needed.

The use of myofunctional therapy (MFT) alone when dealing with pediatric SDB has not been widely investigated. Results of studies performed on children with orthodontic problems have shown that isolated extensive and well-controlled MFT can lead to a return to a normal oral-facial anatomy [31], but the effect on SDB is unknown. The absence of MFT in association with T&A or other treatments to address dysmorphism induced by dysfunctions has been clearly shown to lead to the persistence or recurrence of pediatric SDB.

MFT and proper tongue positioning in the oral cavity have been described since 1918 as leading to an improvement in mandibular growth, nasal breathing, and facial appearance. MFT is comprised of isotonic and isometric exercises that target oral (lip, tongue) and oropharyngeal structures (soft palate, lateral pharyngeal wall). Breathing, particularly nasal breathing, swallowing, mastication, and suction are some of the daily functions that help the oral cavity gain growth during early childhood and participate in the normal development of the oral-facial structures. Normal development of oral-facial structures is important for air exchange, particularly during sleep. During childhood sleep, the tongue will be positioned against the palate and help widen the palate (the adult width is between 40 and 50 mm). The continuous interaction of the tongue with active intermaxillary synchondrosis and the alveolodental growth region are factors in the normal development of oral-facial structures [32]. MFT [32, 33] aims to obtain appropriate head posture, appropriate positioning of the tongue on the palate against the upper teeth, appropriate swallowing, appropriate mastication using both sides and posterior chewing, appropriate breathing through the nose and keeping the mouth closed, appropriate "cleaning" of the nose, and appropriate speech and articulation. Active parental involvement is required to obtain valid results. Specialized educators exist in many countries, but educational programs vary widely in depth. A meta-analysis [33] showed that MFT, in association with other therapeutic approaches, may lead to complete remission of OSA in about 60% of children with whom it is used. The major problem is compliance with daily exercises and continuous parental involvement with the training exercises of the child; this treatment approach is called "active MFT" [33].

A "passive MFT" [34] has been tried recently but results are only available in an abstract form; however, such an approach would be very useful: it calls upon mandibular devices that would lead to sensory stimulation of the tongue, leading to tongue muscle activity. More work in this area is needed.

Positive Airway Pressure

When all of the above fails, or when a syndromic presentation or hypoventilation during sleep is present, positive airway pressure is considered. However, long-term follow-up of children treated with masks that are placed on a developing face reveals a negative effect: the impairment of facial growth, already well documented in children, may occur within 1 year. Furthermore, use of a chin strap that "pushes back the chin due to its anchorage" increases this very negative effect. Specialists have used a mask with pressure on the forehead to decrease this important and ignored negative effect, and have associated positive airway pressure usage with daytime MFT, but currently this negative effect is often poorly addressed.

Conclusion

We have made progress in our understanding of pediatric OSA, and we can even recognize factors leading to its development or worsening. However, pediatricians and pediatric subspecialists are often unaware of the advances and the remedies available. The frequency of pediatric OSA can be significantly decreased if "fundamental functions" (nasal breathing, sucking, swallowing, masticating, and phonation) are regularly evaluated for appropriate development, and if any existing defect is properly addressed. Treatments allowing normal development exist, and educational material and training for parents is available; the education of specialists and diffusion of existing knowledge, including to parents, is needed [35].

Acknowledgment

Studies were supported by Chang Gung Memorial Hospital grants (No. CRRPG5C0171, 172, and 173) to Y.S. Huang.

References

1 Guilleminault C, Eldridge F, Simmons F, Dement WC: Sleep apnea in eight children. Pediatrics 1976;58:23–30.
2 Guilleminault C, Korobkin R, Winkle R: A review of 50 children with obstructive sleep apnea syndrome. Lung 1981;159: 275–287.
3 Guilleminault C, Winkle R, Korobkin R, Simmons B: Children and nocturnal snoring: evaluation of the effects of sleep related respiratory resistive load and daytime functioning. Eur J Pediat 1982;l39:l65–l7l.
4 Redline S, Tishler PV, Schluchter M, Aylor J, Clark K, Graham G: Risk factors for sleep-disordered breathing in children: associations with obesity, race, and respiratory problems. Am J Respir Crit Care Med 1999;159:1527–1532.
5 Schwab RJ, Pasirstein M, Pierson R, Mackley A, Hachadoorian R, Arens R, et al: Identification of upper airway anatomic risk factors for obstructive sleep apnea with volumetric magnetic resonance imaging. Am J Respir Crit Care Med 2003;168:522–530.
6 Ye J: Emerging role of adipose tissue hypoxia in obesity and insulin resistance. Int J Obes 2009;33:54–66.
7 Bonsignore MR, McNicholas WT, Montserrat JM, Eckel J: Adipose tissues in obesity and obstructive sleep apnea. Eur Respir J 2012;39:746–767.
8 Aknin JJ: La croissance cranio-faciale. Paris, SID, 2007, pp 1–267.
9 Guilleminault C, Primeau M, Yuen K, Leger D, Metlaine A: Sleep-disordered breathing in Ehlers-Danlos syndrome: a genetic model of OSA. Chest 2013;144: 1503–1511.
10 Ben-Bassat Y, Brin I: Skeletal and dental patterns in patients with severe congenital absence of teeth. Am J Orthod Dentofac Orthop 2009;135:349–356.
11 Lee SY, Guilleminault C, Chiu HY, Sullivan SS: Mouth breathing, "nasal disuse," and sleep-disordered breathing. Sleep Breath 2015;19:1257–1264.
12 Shimizu T, Maeda T: Prevalence and genetic basis of tooth agenesis. J Dent Sci Rev 2009;45:52–58.
13 Guilleminault C, Cummiskey J, Motta J, Lynne-Davies P: Respiratory and hemodynamic studies during wakefulness and sleep in myotonic dystrophy. Sleep 1978;1:19e31.
14 Harvold EP, Tomer BS, Vargervik K, Chierici G: Primate experiments on oral respiration. Am J Orthod 1981;79:359–372.
15 Vargervik K, Miller AJ, Chierici G, Harvold E, Tomer BS: Morphologic response to changes in neuromuscular patterns experimentally induced by altered modes of respiration. Am J Orthod 1984;85:115–124.
16 de Castro Martinelli R, Queiroz Marchesan I, Berretin Felix G: Longitudinal study of the anatomical characteristics of the lingual frenulum and comparison to literature. Rev CEFAC 2014;16:1202–1207.
17 de Castro Martinelli R, Queiroz Marchesan I, Jordao Gusmao R, de Castro Rodriguez A, Berretin-Felix G: Histological characteristics of altered lingual frenulum. Intern J Pediatr Child Health 2014; 2:5–9.
18 Kotlow LA: Oral diagnosis of oral frenum attachments in neonates and infants: evaluation and treatment of the maxillary and lingual frenum using the Erbium:YAG laser. J Pediatr Dent Care 2004;10:11–13.
19 Ruffoli R, Giambelluca MA, Scavuzzo MC, et al: Ankyloglossia: a morphofunctional investigation in children. Oral Dis 2005;11:170–174.
20 Defabianis P: Ankyloglossia and its influence on maxillary and mandibular development: a seven-year follow-up case report. Funct Orthod 2000;17:25–33.
21 García Pola MJ, González García M, García Martín JM, Gallas M, Seoane Lestón J: A study of pathology associated with short lingual frenulum. ASDC J Dent Child 2002;69:59–62.
22 Huang YS, Guilleminault C: Pediatric sleep apnea and the critical role of oral facial growth: evidence. Front Neurol 2013;3:184.
23 Bhattacharjee R, Kheirandish-Gozal L, Spruyt K, Mitchell RB, Promchiarak J, Simakajornboon N, et al: Adenotonsillectomy outcomes in treatment of obstructive sleep apnea in children: a multicenter retrospective study. Am J Respir Crit Care Med 2010;182:676–683.
24 Guilleminault C, Li K, Quo S, Inouye RN: A prospective study on the surgical outcomes of children with sleep-disordered breathing. Sleep 2004;27:95–100.
25 Tauman R, Gulliver TE, Krishna J, Montgomery-Downs HE, O'Brien LM, Ivanenko A, et al: Persistence of obstructive sleep apnea syndrome in children after adenotonsillectomy. J Pediatr 2006;149:803–808.
26 Guilleminault C, Huang YS, Glamann C, Li K, Chan A: Adenotonsillectomy and obstructive sleep apnea in children: a prospective survey. Otolaryngol Head Neck Surg 2007;136:169–175.
27 Amin R, Anthony L, Somers V, et al: Growth velocity predicts recurrence of sleep-disordered breathing 1 year after adenotonsillectomy. Am J Respir Crit Care Med 2008;177:654–659.
28 Huang YS, Guilleminault C, Lee LA, Lin CH, Hwang FM: Treatment outcomes of adenotonsillectomy for children with obstructive sleep apnea: a prospective longitudinal study. Sleep 2014;37:71–76.

29 Guilleminault C, Abad VC, Chiu HY, Peters BR, Quo S: Missing teeth and pediatric obstructive sleep apnea. Sleep Breath 2016;20:561–568.
30 Huang YS, Quo S, Berkowski JA, Guilleminault C: Short lingual frenulum and obstructive sleep apnea in children. Int J Pediatr Res 2015;1:1–4.
31 Chauvois A, Fournier M, Girardin F: Reeducation des fonctions dans la therapeutique orthodontiques. Paris, SID, 1999.
32 Pirelli P, Saponara M, Guilleminault C: Rapid maxillary expansion in children with obstructive sleep apnea. Sleep 2004;27:761–766.
33 Guilleminault C, Akhtar F: Pediatrics sleep-disordered-breathing: new evidence on its development. Sleep Med Rev 2015;14:1–11.
34 Camacho M, Certal V, Abdullatif J, Zaghi S, Ruoff CM, Capasso R, Kushida C: Myofunctional therapy to treat obstructive sleep apnea: a systematic review and meta-analysis. Sleep 2015;38:669–675.
35 Chuang LC, Lian YC, Hervy-Auboiron M, Guilleminault C, Huang YS: Passive myofunctional therapy applied on children with obstructive sleep apnea: a 6-month follow-up. J Formos Med Assoc 2017, in press.
36 Bahr D: Nobody Ever Told Me (or My Mother) That. Arlington, Sensory World, 2010.

Prof. Christian Guilleminault, MD, Biol D
Stanford University Sleep Medicine Division
450 Broadway Street, Pavilion B, 2nd Floor, MC 5730
Redwood City, CA 94063 (USA)
E-Mail cguil@stanford.edu

Lin H-C (ed): Sleep-Related Breathing Disorders. Adv Otorhinolaryngol. Basel, Karger, 2017, vol 80, pp 145–147
DOI: 10.1159/000470886

Future Perspectives in Sleep Medicine

Leh-Kiong Anne Huon • Christian Guilleminault
Stanford University Sleep Medicine Division, Redwood City, CA, USA

Abstract
"Sleep Medicine" is now a specialty in its own right. Currently, there is increasing recognition of the very negative impact sleep disorders have on learning, education, safety, and quality of life. Technological advances will help us to break down diagnoses (e.g., narcolepsy has now been subdivided into types 1 and 2, depending upon the hypocretin levels in the spinal fluid) and to discover relationships to other bodily systems (e.g., type 1 narcolepsy potentially being an autoimmune disorder). The modern lifestyle of many, as characterized by a shortening of sleep periods, shift work, jet lag, and the need to be constantly available, means that advances in sleep medicine may result in a major understanding of more balanced "work-rest lifestyle" modifications.

Sleep medicine is now a specialty in its own right. Examination for the diploma of Specialist in Sleep Medicine is administered by ECMFG, the official organization controlling internship and residency in the USA. There is a sleep-medicine specialist examination and diploma conducted by the European Sleep Research Society (ESRS), and there is an international examination and diploma administered by the World Association of Sleep Medicine (WASM). There are currently numerous clinical PhD qualifications available in many parts of the world. In 2016, a "World Sleep Society" (WSS) was formed by the joint efforts of WASM and the World Federation of Sleep-Medicine Societies (WFMS). This WSS organizes a congress every 2 years in different parts of the world (hosted in Prague in October 2017) and has specific subgroups with specialized committees, such as the Surgeons and Orthodontists Committee and the Basic Sleep Research Committee. The pediatric sleep medicine field is not only represented at the congress of the WSS, but has its International Pediatric Sleep Association (IPSA) congress in the years when the WSS has no congress. There are continental societies such as the Latin America Sleep Society, the European Sleep Research Society (ESRS), the Asian Sleep Society, the Indian Sleep Society, and the American Association of Sleep Medicine (AASM), with its research associate the "Sleep Research Society." These different associative groups perform both an educational and political role, dealing with national and international agencies to integrate sleep medicine into the curricula of medical schools all over the world, to have patients with sleep disorders treated by competent specialists, and to achieve national insurance coverage for all patients and specialists involved in their care.

Clear discrepancies exist depending on the part of the world one looks at. However, all parts of the world have an interest in sleep and a progressive understanding of the very negative impact that sleep disorders have on learning, education, safety (from road safety to industrial safety), the quality of life of the general population, and worsening of the recognized first 3 causes of early death: accidents, cancer, and cardiovascular events. Governments are finally appreciating that sleep disorders have a major impact on the "growth national product" and "happiness index" of their citizens. Advances clearly have to be made in all these aspects, but recognition of the problem and documentation of negative impacts and the trajectory to remedy the situation exist and are available to all.

National Health providers are finally realizing the socioeconomic impact of "sleep disorders." The challenge will be to respond to the problem without the current fear of major payers restricting treatments and creating rules such as the "Medicare rules" in the USA that deny appropriate care to a clear portion of the general population, and in particularly to women.

There have been many advances in the field of research and more are to come. As an example, narcolepsy, which is now subdivided into type 1, with abnormally low cerebrospinal fluid-hypocretin levels (and usually cataplexy), and type 2, which lacks this key symptom, is being researched today as a potential model of autoimmune disorders of the brain. The demonstration of the association between HLA DQB1*0602 with a specific polymorphism of the T cell receptor alpha in narcolepsy type 1, along with absent or pathologically low levels of hypocretin in cerebrospinal fluid, has led researchers in this direction. It indicates that a specific, well-defined sleep disorder will allow us to understand further interactions between immune factors and brain functioning and vulnerability.

The problem of "sleepiness" has led to sophisticated studies of brain functioning using advanced imaging technologies looking at the impact on learning, memory creation, and cognition. These important functions, which are fundamental to mankind and allow humanity to progress, have been linked to sleep and its 2 states: rapid eye movement and non-rapid eye movement. Investigations of disturbances to sleep and the sleep-wake cycle and its recoveries have also opened a new window in our understanding of the plasticity of the brain, and its possible recovery from major insults, such as in traumatic brain injury.

The interaction between sleep restriction and the controls of many peptides, including those involved in hunger and satiety feelings, have allowed a better understanding of the development of metabolic disorders, including obesity and diabetes, hopefully opening an improved option to new therapies. The shortening of sleep, the creation of uncontrolled shift work, regular jet lags, the demands of the cell phone, and the globalization of economies have been demonstrated to have an important role in the development of cancers, particularly breast cancer in women and prostate cancer in men. However, new research is looking at more generalized phenomena, such as the impairment of DNA-RNA messaging and creation of more errors at the cellular level that overload the capabilities of T cells to eliminate abnormally created cells. Similarly, the impact of the change in sleep-wake cycles and its impact on specific genes, such as ApoE-4, which may predispose to an abnormal degeneration of brain cells and lead to neurodegenerative disorders, is another field of research.

Specific sleep disorders that are known to all sleep specialists, such as sleep-disordered breathing, have attracted the attention of different health specialists, including orthodontists and oral surgeons, as well as geneticists. Sleep-disordered breathing "runs" in families, indicating the presence of familial risk factors, and addressing these early in life may avoid treatment at a later stage. Many questions prevail. Is the "obesity gene" more prevalent in some ethnicities, leading to obesity hypoventilation early in life or obesity with comorbid obstructive sleep apnea, and its cohort of cardiovascular complications? Are there functional changes that occur early in life that modify the trajectory of the buildup of the oral-facial region, particularly during childhood, increasing the risk of collapsibility of the upper airway during sleep? How can we recognize these changes as early as possible and address them before they impact on sleep-related reflexes? Are these functional changes more common in a specific genetic background and what impact does fetal life have of the enhancement of these risk factors, either functionally or genetically?

Currently, we often retain a dated perspective, with surgery being the only solution as the syndrome is too involved. If surgery on soft tissues of the upper airway is clearly not the solution, modification of the bone structure is considered as a possibility, from surgically assisted rapid maxillary expansion to maxilla-mandibular surgery, or to the 2 combined, as 1 type of surgery does not completely address the 3-dimensional abnormal growth noted in many of these patients. These surgeries indicate the failure of early recognition and early intervention. They have their current role, but the goal is prevention and early recognition and treatment. This will involve genetic and fetal research, but also the contributions of obstetricians, pediatricians, pediatric ENT, pediatric dentists, and orthodontists. Some countries, such as Brazil, have made it mandatory at birth to check the lingual and nasal frenulum in the same way that vision and audition are checked. Clearly, a short lingual frenulum is a phenotype rarely checked in many countries by individuals dealing with sleep-disordered breathing. We have aimed at surgery and nerve stimulation approaches studied since 1989, such as hypoglossal stimulation. Between 1995 and 2005, hypoglossal stimulation led to a greater understanding of the different nerve fibers involved in the different activities of the tongue, for example from protrusion to retraction, but did not lead to a better treatment of the problem. Engineers have been able to better define the frequencies needed to stimulate the majority of protrusion fibers, but diffusions of stimulation exist and variable results will always be seen.

The ultimate goal is prevention and early recognition in order to avoid having to deal not only with the problem, but also its consequences at any age, from cardiovascular comorbidities to nasal disuse and the need to reeducate the abnormal function. Future research will flourish. It is known that there is a relationship between inflammation and the sleep-wake cycle, and that inflammatory cytokines are secreted in association with poor sleep, and such inflammatory factors may be the pathways through which syndromes may propagate. Now that we know there are direct connections from lymph canals to the brain [1], we may even better understand the brain impairments associated with sleep disorders. Sleep medicine specialists and researchers have very large avenues of research and their discoveries and advances will touch all fields of medicine.

Additional Resources

World Association of Sleep Medicine: www.wasmonline.org
World Sleep Society: www.worldsleepsociety.org
World Sleep Federation: www.worldsleepfederation.org
European Sleep and Research Society: www.esrs.eu
American Academy of Sleep Medicine: www.aasm.net.org

Reference

1 Antoine Louveau A, Smirnov I, Keyes TJ, et al: Structural and functional features of central nervous system lymphatic vessels. Nature 2015;523:337–341.

Prof. Christian Guilleminault, MD, Biol D
Stanford University Sleep Medicine Division
450 Broadway Street, Pavilion B, 2nd Floor, MC 5730
Redwood City, CA 94063 (USA)
E-Mail cguil@stanford.edu

Lin H-C (ed): Sleep-Related Breathing Disorders. Adv Otorhinolaryngol. Basel, Karger, 2017, vol 80, pp 148–154
DOI: 10.1159/000470887

Future Perspectives in Sleep Surgery

Michael Friedman[a, b] • Anna M. Salapatas[b] • Lauren B. Bonzelaar[c]

[a]Section of Sleep Surgery, Rush University Medical Center, and [b]Section of Otolaryngology, Chicago ENT, Advanced Center for Specialty Care, Advocate Illinois Masonic Medical Center, Chicago, IL, and [c]Department of Anesthesiology, Loyola University Medical Center, Maywood, IL, USA

Abstract

Over the past few decades there has been a significant increase in research focusing on the pathophysiology, diagnosis, and treatment of sleep apnea. However, there is still a long way to go in creating standard evidence-based medical practice for the diagnosis, evaluation, and treatment of sleep apnea patients. Current and future directions of sleep surgery and other treatments are discussed here.

Surgical treatment for obstructive sleep apnea/hypopnea syndrome (OSAHS) is one of the most controversial and challenging, yet ever-evolving specialties that otolaryngologists perform. While many sleep medicine specialists deny the role of surgery as a treatment option, it is clear that many patients with OSAHS cannot or will not accept continuous positive airway pressure (CPAP) therapy, even though it can bypass upper airway obstruction. Sleep surgeons, however, continue to introduce many new and innovative techniques that yield desirable results in selected patients. As the treatment of sleep disorders has expanded to nearly every health-related specialty, the role of the otolaryngologist is undoubtedly the most vital, offering the most comprehensive management of OSAHS, including physical examination, diagnostic testing, and nonsurgical and surgical treatment.

Over the past few decades, there has been a significant increase in research focusing on the pathophysiology, diagnosis, and surgical treatment of OSAHS. This has allowed sleep surgeons to recognize factors that direct decision making for appropriate treatment, and offer various alternatives to reduce or possibly cure symptoms. The future of sleep medicine will rely on the early recognition of OSAHS symptoms and selection of the best treatment for each patient to promote more desirable outcomes. Although great strides have been made in understanding the treatment of this disease, there are many areas of the field that remain controversial or show inconsistent findings. While most fields of medicine focus on returning patients to a "normal" or healthy state, the challenging aspect of sleep medicine resides in the fact that some sleep apnea patients are anatomically or genetically predisposed to upper airway obstruction, and their diseased state is actually their "normal" state. Therefore, treatment is based on altering their natural anatomy.

Developing both invasive and noninvasive methods for treatment is crucial for the progress of the field. It is important to remember that surgery is only a salvage approach for patients who have failed noninvasive treatment. A surgeon whose expertise includes both invasive and noninvasive options to cure or palliate symptoms is in the ideal position to offer the most comprehensive treatment for OSAHS. However, these 2 treatment methods are not mutually exclusive. Some patients may require an invasive treatment to enhance the efficacy of their long-term, noninvasive alternative.

Why Is There a Future Role for Sleep Surgery in View of Continuous Positive Airway Pressure Success?

Sleep surgery is currently, and will likely always be, a suboptimal alternative to CPAP. In spite of this, when curative or standard therapy is not an option, the palliative control of symptoms must be offered to patients. While arguments against sleep surgery are based on the low rate of surgical cures, most diseases that we treat have low or nonexistent cure rates. Hypertension, diabetes, morbid obesity, and asthma are almost never cured. We do not, however, deny palliative treatments to alleviate symptomatology and reduce the risk of long-term sequelae.

Despite continuous improvements, CPAP compliance rates are not likely to reach 100%. More importantly, CPAP compliance studies are performed on patients who have symptoms severe enough to accept diagnostic polysomnography (PSG) and CPAP trials. A large percentage, and perhaps the majority, of patients with OSAHS have mild disease. Most of these patients defer testing and diagnosis because they are reluctant to accept CPAP therapy. Sleep surgery may be a beneficial and important treatment option for these patients. Even when surgery is less effective than CPAP, we should present all options and allow our patients to make informed decisions.

Diagnosis

For the identification of disease and as an indication of treatment efficacy, nearly every physician or institution has their own methods of screening. Classical medical treatment is based on 3 critical points: history, physical exam, and diagnostic testing. The diagnosis of OSAHS has been based on symptomatology and PSG findings. A patient presenting with symptoms suggestive of OSAHS should be promptly sent for confirmatory PSG with the immediate institution of CPAP therapy. Since OSAHS is a disease of sleep, patients who sleep alone will not be aware of symptoms unless they are at a late stage with significant complications, and patients who are examined for another pathology are not likely to volunteer information about snoring or daytime somnolence. Even patients with bed partners are often in great denial about sleep apnea symptoms. Thus, the history obtained from a bed partner or family member is often essential for an accurate assessment.

In medicine, the identification of disease often requires an early recognition of physical findings before symptoms present or when patients provide a poor history. While often ignored by medical sleep specialists, physical examination of the airway is crucial for the sleep surgeon. As otolaryngologists, we are in the unique position of examining and assessing the upper airway in every patient. In this role, we can identify patients with an unfavorable anatomy who are likely to have OSAHS, even without a complaint of snoring. The neck circumference, Friedman tongue position (FTP), tonsil size, and lingual tonsil grade should always be noted in every patient. Alerting patients to the potential existence of OSAHS will often help them seek information from a bed partner or roommate, and eventually result in a diagnosis.

Once a clinical identification of OSAHS symptoms has been established, PSG remains the most accepted diagnostic tool for the presence and severity of the disease. There are, however, various limitations associated with PSG testing; 1 major caveat is the variability in the definition of hypopnea seen in different laboratories. Because hypopnea is a component of the apnea-hypopnea index (AHI), the cardinal measurement of PSG, it is important to have consistency in its definition. Many sleep surgeons also question the ability of PSG to accurately describe the disease and its severity, noting that AHI does not always correlate with clinical symptoms. Since patients seek treatment for the clinical manifestations of sleep apnea, including snoring, daytime fatigue, and long-term health consequences, it is important to diagnose severity in relation to those symptoms. Other tests, such as the multiple sleep latency test, psychomotor vigilance test, and maintenance of wakefulness test, are also used in diagnosis. However, these tests focus on symptomatology, and therefore cannot be compared to one another. A common language needs to be established to improve our ability to compare studies, determine trends, and assess outcomes.

Many patient questionnaires or surveys, as well as some objective physical measurements, have been suggested to predict the presence of OSAHS; however, a consensus is far from established. Although PSG is still the gold standard for diagnosis and determining the severity of OSAHS, validated quality of life questionnaires are now recognized as important tools for measuring treatment efficacy. The most commonly used questionnaires include the Epworth Sleepiness Scale, Functional Outcomes of Sleep Questionnaire, Sleep Apnea Quality of Life Index, and Pittsburgh Sleep Quality Index. Since quality of life is the most important aspect for patients, symptom reduction is ultimately the most essential component of surgical treatment, regardless of obtaining objective success in AHI outcomes.

Evaluation of treatment success should be made on an individual basis rather than relying on 1 test, or set of tests. Furthermore, individual comorbidities, symptoms, and concerns should be taken into consideration when determining if a patient has been cured. Many factors play a role in these outcome measures. The goal of sleep surgery must always be a combination of AHI reduction with the improvement or elimination of symptoms. Achieving either goal is meritorious but incomplete. Sleep surgeons may see poor success in AHI reduction; however, surgery has been shown to be effective in improving quality of life for the majority of patients.

Surgical planning requires a good understanding of the patient's upper airway anatomy. Nasal patency is important, and a standard ENT examination will identify a deviated septum, turbinate hypertrophy, polyps, or adenoid tissue obstructing the airway. Endoscopic assessment is now the standard of care in the clinic, and will likely replace anterior rhinoscopy altogether. Nasal valve collapse is a common problem and may be more frequent in patients with OSAHS, since oropharyngeal or hypopharyngeal collapse results in increased negative pressure within the nasal ala. The identification and correction of a collapse should be an important part of diagnosis and treatment.

Staging

The oropharynx and hypopharynx are first assessed with the Friedman anatomic staging system [see the paper by Friedman et al., this vol., pp. 41–48]. Lingual tonsil hypertrophy (LTH) is best assessed with the patient awake and with different tongue positions. The LTH grade should be established in every OSAHS patient [see the paper by Friedman et al., this vol., pp. 41–48]. Since surgical treatment options may affect the tongue size and position, the effect of tongue protrusion on the hypopharyngeal airway should be assessed, both while the patient is awake and, if feasible, asleep.

Clinical screening with FTP allows evaluation of the patient with a simple examination. Patients with a favorable anatomy (FTP I or II and large tonsils), even with severe disease, will often be directed toward surgery if CPAP is not accepted. Patients with severe disease and an unfavorable anatomy (FTP III or IV) will be directed away from surgery until all noninvasive options have been exhausted. Patients with mild disease and CPAP failure who have an unfavorable anatomy can be directed to mandibular advancement therapy.

Anatomic staging is simple and should be the initial part of every patient's treatment plan. The routine and concurrent practice of LTH grading can provide a similar insight into which patients will benefit from lingual tonsillectomy. Additionally, LTH grading provides a universal language with which to identify the presence and grade of lingual tonsil tissue, and a basis on which to measure surgical outcomes of lingual tonsillectomy. Future studies are needed to determine which LTH grades benefit most from surgery.

Drug-Induced Sleep Endoscopy

Once a patient is considered a candidate for surgical treatment of OSAHS, drug-induced sleep endoscopy (DISE) will help identify targeted areas requiring treatment. The evaluation necessitates the pharmacologic induction of sedation and flexible fiber-optic endoscopy to visualize the upper airway obstruction and/or snoring. DISE uniquely offers a dynamic evaluation of the upper airway during conditions that ideally mimic natural sleep, and also enables the visualization of specific structures that contribute to upper airway obstruction.

Having already established the need for surgical options, future goals will aim at perfecting the technique and selection criteria to maximize successful outcomes. The key points are: (1) recognition that OSAHS, whether mild or severe, is often caused by multiple levels of obstruction, and therefore treatment should aim at multilevel correction, (2) the risks and morbidity of surgery must be consistent with the severity of the disease, and (3) the future will offer more minimally invasive options for mild disease and more maximally invasive options for severe disease.

Palatal Procedures

Although classic uvulopalatopharyngoplasty (UPPP) is still used as a basic surgical technique for mild OSAHS, it is likely to become an extremely rare procedure for mild disease in the future. The morbidity of classic UPPP, which includes uvulectomy, is often considered unacceptable since less invasive procedures are available. Palatal stiffening with pillar implants or the injection of sclerotherapy agents will continue to offer a simple option for patients whose major complaints are snoring. New techniques of palatal suturing, which scar the palate in an anterior position, are likely to have continuous improvement and become popular for patients with mild disease or snoring. Devices such as Zelegent (formerly known as Snorx) have been developed as an clinic-based outpatient approach to shortening the soft palate.

The next level of treatment is the lateral pharyngoplasty [1, 2], which will replace classic UPPP as the standard workhorse for the treatment of mild to moderate OSAHS. This procedure involves surgically splinting the lateral pharyngeal walls, achieved by turning the palatopharyngeus muscles into dilating flaps. Good candidates for the procedure are those who have a clearly defined, bulky posterior tonsillar pillars (palatopharyngeus muscle), and a soft palate that does not touch the posterior pharyngeal wall while the subject is sitting upright and awake. The longer the distance between the soft palate and the tongue, the easier the surgical approach to the superior pharyngeal constrictor muscles. Lateral pharyngoplasty requires general anesthesia and has significant postoperative pain, but retains the uvula and has a lower rate of morbidity than classic UPPP.

Severe disease will require invasive multilevel treatment. Palatal reconstruction with additional modified procedures, such as Z-palatoplasty, may play a role. Specific patients with a relatively long hard palate and a short soft palate will benefit from transpalatal advancement pharyngoplasty [3], which alters the retropalatal airway by advancing the palate forward. Transpalatal advancement pharyngoplasty offers a potential alternative to aggressive UPPP in an attempt to enlarge the upper oropharyngeal airway and improve respiratory function. This advancement affects both the upper and lateral pharyngeal airway. Many other manifestations of palatal surgery with uvular preservation will continue to evolve.

When tonsillectomy is planned, coblation, ultrasound, and tools other than cold steel have become popular options. However, studies have not proven any significant benefit over cold steel for total tonsillectomy. Subtotal or intracapsular tonsillectomy may be adequate and resolve airway obstruction in some patients. Coblation or radiofrequency reduction of large tonsils in these patients has been proven to be less painful and have a reduced rate of postoperative bleeding.

Hypopharyngeal Treatment

Hypopharyngeal obstruction remains one of the greatest challenges in OSAHS treatment. The future is likely to bring techniques of submucosal tongue base reduction without mucosal injury. Currently, this is available only with radiofrequency, which has proven to be safe but with a limited ability to shrink tissue. Radiofrequency treatment of the tongue will continue to be part of multilevel treatment planning for patients with mild disease.

Patients with more severe disease will require more aggressive tissue reduction. Coblation techniques and modification of the submucosal minimally invasive lingual excision (SMILE) technique [4] will continue to be an important part of multilevel treatment for patients with moderate to severe OSAHS. The advantage of this technique is that the bulk removal of lingual tissue can be performed under oral ultrasound and endoscopic guidance, which leaves the mucosal surface intact. This significantly decreases postoperative morbidities often associated with techniques that are destructive to the mucosa, and allows the aggressive reduction of the macroglossia with minimal risk to tongue function. Coblation (radiofrequency) technology is also capable of reducing LTH.

Transoral robotic surgery (TORS) has been established as another novel technique [5] for maximal tongue base reduction and lingual tonsillectomy. TORS was initially used for the resection of base-of-tongue neoplasms in 2006, as it provided improved magnification, surgical precision, avoidance of critical vessels and nerves, hemostatic control, and the meticulous handling of tissues. Studies observing OSAHS outcomes of TORS for tongue base and lingual tonsil resection provided remarkable functional results, including minimal pain, preserved swallowing, and improved quality of life. The indication of TORS for lingual tonsillectomy, however, will require further study. While logic would dictate that lingual tonsillectomy is beneficial for patients with LTH grades 3 and 4, it is not clear that the morbidity of lingual tonsillectomy is justifiable for LTH grade 1 or 2. To date, no study has proven the value of lingual tonsillectomy for OSAHS. Furthermore, the advantages of TORS come at the expense of longer total operative times and greater surgical costs. The cost of purchasing the robot and the associated maintenance expenses must also be considered. Nevertheless, PSG results and complication rates have been encouraging for TORS as part of multilevel surgery for OSAHS.

Suspension and Expansion

Tongue base and hyoid suspension procedures, such as genioglossal advancement, tongue base stabilization, and hyothyroid or hyomandibular suspension will play a minor role in selected patients. Bimaxillary advancement has always been considered the ultimate treatment option for OSAHS. Although success rates (not cure) have been reported to be as high as 90%, morbidity is significant. Bimaxillary advancement will always play a limited role in the treatment of OSAHS.

Nonsurgical maxillary expansion should be an important consideration in future OSAHS treatment. The early identification of residual OSAHS after tonsillectomy in pediatric patients allows for nonsurgical expansion. Chronic manifestations of OSAHS can be deleterious to child development, and thus early recognition and prompt treatment are crucial in the pediatric population. Although tonsillectomy and adenoidectomy improve symptoms in most pediatric cases, the procedure has been ineffective in fully ameliorating disease in some patients. Without postoperative PSG, we may miss an opportunity to offer adjunctive treatment before full growth is attained. Oral devices for maxillary expansion have been used to treat OSAHS among children. It is believed that rapid maxillary expansion decreases nasal resistance and facilitates the passage of air through the nose [6], thus reducing mouth breathing. Additionally, it increases the maxillary dental arch and improves the position of the tongue, which enables proper sealing of the lips when the mouth is closed.

This form of skeletal expansion can also be proposed to address adult maxillary hypoplasia in the setting of OSAHS. Surgical mini-implant-assisted rapid maxillary expansion utilizes anchored implants adjacent to the mid-palatal suture [7]. The principles of distraction osteogenesis are applied even for bone growth along the hard palate. This results in the expansion of the nasal floor and widening of the posterior hard palate, which is intimately associated with the airway dilator muscles.

One of the most promising treatments for the future will be hypoglossal nerve stimulation therapy [8–10]. In contrast to other surgical procedures that directly alter the upper airway anatomy to prevent obstruction, hypoglossal nerve stimulation reduces obstruction by increasing the glossal muscle tone. This targeted stimulation prevents the tongue from falling back and occluding the airway during sleep. Early studies have demonstrated a 60% success rate in selected patients. The number of adverse events associated with the therapy continue to be acceptable, with the majority being device-related events occurring in association with device implantation. At present, the therapy is limited to patients with a BMI <32 and an AHI <50. Patients with a large soft palate, large faucial or lingual tonsils, or concentric collapse seen during DISE are not candidates. Future research, including optimization of device features and stimulation parameters as well as patient selection, is necessary to make hypoglossal nerve stimulation a viable alternative to positive airway pressure therapy and upper airway surgical procedures.

Conclusion

Surgery will continue to be a vital component of treatment for OSAHS. A certain percentage of patients will always be intolerant of CPAP and will benefit from surgery either for the cure or palliation of symptoms. Surgery is frequently discounted because it often fails to bring the AHI below 15. However, AHI does not always correlate with other clinical values associated with OSAHS. In spite of a relatively poor success in AHI reduction, surgery has been shown to be effective in improving quality of life for the majority of patients. If surgery is reserved for patients who cannot accept CPAP therapy, this salvage treatment should be viewed as a valuable alternative that can alleviate symptoms and improve quality of life. Further improvement in surgical success will depend on bettering our selection of the appropriate surgical procedure for each patient. Continued advancement of these techniques targeting each anatomic area of obstruction will reduce morbidity and augment results.

References

1 Pang KP, Woodson BT: Expansion sphincter pharyngoplasty: a new technique for the treatment of obstructive sleep apnea. Otolaryngol Head Neck Surg 2007;137:110–114.

2 Cahali MB: Lateral pharyngoplasty: a new treatment for obstructive sleep apnea hypopnea syndrome. Laryngoscope 2003;113:1961–1968.

3 Woodson BT: Retropalatal airway characteristics in uvulopalatopharyngoplasty compared with transpalatal advancement pharyngoplasty. Laryngoscope 1997;107:735–740.

4 Maturo SC, Mair EA: Submucosal minimally invasive lingual excision: an effective, novel surgery for pediatric tongue base reduction. Ann Otol Rhinol Laryngol 2006;115:624–630.

5 Justin GA, Chang ET, Camacho M, Brietzke SE: Transoral robotic surgery for obstructive sleep apnea: a systematic review and meta-analysis. Otolaryngol Head Neck Surg 2016;154:835–846.

6 Machado-Júnior AJ, Zancanella E, Crespo AN: Rapid maxillary expansion and obstructive sleep apnea: a review and meta-analysis. Med Oral Patol Oral Cir Bucal 2016;21:e465–e469.

7 Cureton SL, Cuenin M: Surgically assisted rapid maxillary expansion: orthodontic preparation for clinical success. Am J Orthod Dentofacial Orthop 1999; 116:46–59.

8 Woodson BT, Soose RJ, Gillespie MB, Strohl KP, Maurer JT, de Vries N, Steward DL, Baskin JZ, Badr MS, Lin H, Padhya TA, Mickelson S, Anderson WM, Vanderveken OM, Strollo PJ Jr; STAR Trial Investigators: Three-year outcomes of cranial nerve stimulation for obstructive sleep apnea: The STAR trial. Otolaryngol Head Neck Surg 2016;154: 181–188.

9 Schwartz AR, Bennett ML, Smith PL, De Backer W, Hedner J, Boudewyns A, Van de Heyning P, Ejnell H, Hochban W, Knaack L, Podszus T, Penzel T, Peter JH, Goding GS, Erickson DJ, Testerman R, Ottenhoff F, Eisele DW: Therapeutic electrical stimulation of the hypoglossal nerve in obstructive sleep apnea. Arch Otolaryngol Head Neck Surg 2001;127: 1216–1223.

10 Friedman M, Jacobowitz O, Hwang MS, Bergler W, Fietze I, Rombaux P, Mwenge GB, Yalamanchali S, Campana J, Maurer JT: Targeted hypoglossal nerve stimulation for the treatment of obstructive sleep apnea: six-month results. Laryngoscope 2016;126:2618–2623.

Michael Friedman, MD
ChicagoENT
30 North Michigan Avenue, Suite 1107
Chicago, IL 60602 (USA)
E-Mail mfriedman@chicagoent.com

Author Index

Subject Index